Introduction to
Managing Technology

ENGINEERING MANAGEMENT SERIES

Series Editor: **Dr. John A. Brandon**
                   University of Wales, Cardiff, UK

*Forthcoming

Introduction to Managing Technology

M. W. Cardullo, P.E.

Northern Virginia Graduate Center
Virginia Polytechnic Institute and State University, USA

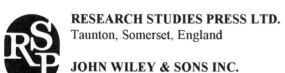

RESEARCH STUDIES PRESS LTD.
Taunton, Somerset, England

JOHN WILEY & SONS INC.
New York · Chichester · Toronto · Brisbane · Singapore

RESEARCH STUDIES PRESS LTD.
24 Belvedere Road, Taunton, Somerset, England TA1 1HD

Copyright © 1996, by Research Studies Press Ltd.
wkp

Marketing and Distribution:

Australia and New Zealand:
Jacaranda Wiley Ltd.
GPO Box 859, Brisbane, Queensland 4001, Australia
Canada:
JOHN WILEY & SONS CANADA LIMITED
22 Worcester Road, Rexdale, Ontario, Canada

Europe, Africa, Middle East and Japan:
JOHN WILEY & SONS LIMITED
Baffins Lane, Chichester, West Sussex, UK, PO19 1UD

North and South America:
JOHN WILEY & SONS INC.
605 Third Avenue, New York, NY 10158, USA

South East Asia:
JOHN WILEY & SONS (SEA) PTE LTD.
37 Jalan Pemimpin 05-04
Block B Union Industrial Building, Singapore 2057

Library of Congress Cataloging-in-Publication Data
Cardullo, M. W. (Mario W.), 1935-
    Introduction to managing technology / M.W. Cardullo.
        p.   cm. - - (Engineering management series ; 4)
    Includes bibliographical references and index.
    ISBN 0-86380-204-4 (Research Studies : Press alk. paper). - - ISBN
0-471-96787-4 (Wiley : alk. paper)
        1. Technology - - Management.   I. Title.   II. Series.
    T49.5.C35   1996
    658.4'062 - - dc20   96-31564
                                                              CIP

British Library Cataloguing in Publication Data
A catalogue record for this book is available from the British Library.

ISBN 0 86380 204 4 (Research Studies Press Ltd.) *[Identifies the book for orders except in America.]*
ISBN 0 471 96787 4 (John Wiley & Sons Inc.) *[Identifies the book for orders in USA.]*

Printed in Great Britain by SRP Ltd., Exeter

To my wife, friend and partner
Karen M. Cardullo
and
my children, Pamela Cardullo Ortiz, Mark, Caroline and Paul Cardullo, but
especially to my grandchildren
Sofiá Marta Cardullo Ortiz
and
Christopher Anthony Cardullo
the future

PREFACE

"...the laws of science do not distinguish between the forward and backward directions of time. However, there are at least three arrows of time that do distinguish the past from the future. They are the thermodynamic arrow, the direction of time in which disorder increases; the psychological arrow, the direction of time in which we remember the past and not the future; and the cosmological arrow, the direction of time in which the universe expands rather than contracts."

Stephen W. Hawking
A Brief History of Time

The management of technology is concerned with all the aspects of time, because the *technological arrow* always points forward. This book is the result of the development of a graduate course of study at Northern Virginia Graduate Center of Virginia Polytechnic Institute and State University ("Virginia Tech"). Many scientists, engineers, and technology managers evolve into the management of technology without formal study of the subject. While practical experience is still an important element in management, understanding is growing on the elements which lead to successful technological development. Numerous academic centers world wide are focusing on training technology managers in these new understandings. This text is designed to serve as an introduction for a course of study in the management of technology ("MOT").

Technological developments are proceeding at an ever increasing pace, forcing enterprises, leadership teams, and others to rapidly adapt. This forward motion requires an understanding of management elements which are necessary to bring technological developments from initial innovation through development and final use within an intended environment. These management elements are not only technical, but have many psychological and sociological components. Coupled with these management elements is the need to understand the historical, or as Professor Hawking calls it, the *psychological arrow*.

Technology does not arise without a history or connection to prior developments. Humanity has been innovating since the beginning of its history and there are lessons to be learned from the successes, and the mistakes in these developments. This human history has also led us to see the need for co-operation between individuals, enterprises and cultures if success is to be achieved. Without group co-operation and communication early societies could not have produced cooper and bronze implements needed for greater productivity. Communication within the management of technology process has been and is an important element to the success of developments in the past, today and in the future.

The development of technology has become a multicultural activity. Technological developments may start in a western laboratory, move to an Asian manufacturer, and be marketed in a developing African nation. The standard course of study for most scientists, engineers, and technologists rarely prepares them for multicultural developments.

The elements of management of technology are presented in this text in a concise form to serve as an introduction to the study to the subject. Each chapter of this book can be and has in some instances been expanded into academic courses and texts. Institutions have realized that engineers, scientists, and technologists need training beyond the classical engineering and science curriculum and encompass current management concepts. The twenty-first century will require the manager of technology to have a global understanding, if technology is to achieve its multifaceted objectives.

Like any development, this text had assistance in moving from conception to a final product. I would like to thank my colleagues at the Northern Virginia Graduate Center ("NVGC") of Virginia Polytechnic Institution and State University ("Virginia Tech"). The idea for the course, which is the basis for this text, started with discussions between the author and Associate Professor Ken Harmon, Director of Industrial and Systems Engineering at NVGC and his helpful suggestions and encouragement. I owe a great deal of gratitude to my associate Dr. C. Howard Robins, Jr. who assisted in editing the course notes which served as the basis for this text. I would like to thank my colleague Dr. Kostas Triantis who reviewed various course notes and made suggestions which improved the material substantially. I would like to especially thank my wife, friend and partner Karen Mandeville Cardullo who read each chapter offering suggestions. If this text is truly readable it is due to her suggestions.

I hope this book will stimulate the reader to continue to investigate the topics. Research on the elements of management of technology are proceeding throughout the world. The results of this research and practical information will serve the reader in expanding upon the issues and topics covered by this text. The results of this expansion of knowledge will lead to technological developments in which society is the final beneficiary.

Mario W. Cardullo

Alexandria, Virginia,
United States of America
June 9, 1996

CONTENTS

CHAPTER 5
Technological Life Cycles and Decision Making

CHAPTER 6
Enterprise Structure and Design

CHAPTER 7
Technology Transfer

CHAPTER 1

Technological Advancement and Competitive Advantage

1. INTRODUCTION

WHY TECHNOLOGY?

The best place to initiate this Socratic quest is in the basis of the word *technology*.

Technology is a general term for the processes by which human beings fashion tools and machines to increase their control and understanding of the material environment (Microsoft 1995). The term is derived from the Greek words *tekhnë*, which refers to an art or craft, and *loggia*, meaning an area of study; thus, technology means the study, or science, of crafting.

Technology consists of knowledge, actions, and accouterments. The principal objective of this text is to present an approach proceeding from knowledge to use technological advancements that meets both societal and competitive needs of the enterprises.

Technological innovations, on a global scale, seem to appear at a rate that increases geometrically, without respect to geographical limits or political systems. However, politics play an important environmental role in technology management. Innovations tend to transform traditional cultural systems, frequently with unexpected social consequences. Thus technology can be perceived as both a creative and a destructive process.

Technology has been a dialectical and cumulative process at the center of human experience. Warnings on the duality of technology, i.e., beneficial and destructive qualities, were observed in the 1950s. It was observed that many products of technology had both useful and harmful or destructive aspects. However, it has been very difficult in practice to predict secondary effects of new technologies. Technology has always been a major means for creating new physical and human environments and one of the principal drivers has been societal needs.

Another of the drivers of technology has been competitive needs of enterprises, both public and private. During the 1980s and 1990s the United States went from the country with the largest trade surplus to one with one of the world's greatest deficits. Prior to the 1980s the United States led the world in competitive advantage, which was secured by its leadership in technology. There is a definite relationship between technology, technological advancements, national and enterprise competitive advantage. Thus we can state that it appears that technology exists due to both a response to societal and competitive needs within nations and enterprises.

1.1 Science and Technological Change

According to Einstein:

> "Der Herrgott ist raffinient aben boshaft ist Er nicht" .... "The Lord is subtle, but he isn't simply mean."

This statement is important for the performance of fundamental research (Jantsch 1961, p. 53). Technological change does not just materialize without being managed. Technological change is driven by enlightened self-interest of either private individuals, organizations, governmental bodies or a combination of these agents of change.

1.2 Breakthroughs

The definition of a breakthrough according to the Concise Oxford Dictionary is:

> "a major advance or discovery....an act of breaking through an obstacle etc."

According to Martino (Martino 1993, p. 187) a technological breakthrough is:

> "An advance in the level of performance of some class of devices or techniques perhaps based on previously utilized principles, that significantly transcends the limits of prior devices or techniques."

This definition infers that adopting a successor technique has a level of inherent capability higher than the prior technology that is considered a breakthrough. Many of the *breakthroughs* that helped shape the twentieth century had a substantial prior history. Einstein stated that he *"..stood on the shoulders of giants."* The following is an example of cascaded events that led to nuclear weapons and atomic energy and now to a major new industry - - *environmental remediation.*

An example of a series of precursor events can be demonstrated by the development of Field Emission Display ("FED") technology. The high cost of existing display technology, combined with a dominant market by a competitor, in this case Japanese manufacturers, led to the development of lighter, less expensive, lower power flat panels for notebook computers. Prior to the development of FEDs, the most popular flat display technology was Liquid Crystal Display ("LCD") or Active Matrix Liquid Crystal Display ("AMLCD") While the LCD is simple to manufacture, and is a mature technology, it can only produce a monochrome display, i.e., black and white. The AMLCD, while capable of producing color, requires a sequential manufacturing process. This type of process requires the completing of many steps before assurance that the final product meets the requirements. The Japanese display manufacturers spent over eight billion dollars during the early 1990s and were not able to reduce the cost of a display below several hundreds of dollars.

In the United States, a consortium was established with Texas Instruments, Motorola, Futaba and Raytheon. The objective of this consortium was to recapture the market for flat panel displays from the Japanese. This consortium developed FED technology. FED is a microtip phosphorus based emission technology that can be described as a very flat Cathode Ray Tube ("CRT") In this technology, electronics are excited off a microtip and bombard a front panel that is coated with Red-Blue-Green ("RBG") phosphors and thus excitement and emission occurs. The technology has been demonstrated in the six inch diagonal size. This technology will serve as a means to help break some of the cost barriers that have been established by the AMLCD industry. Another advance to the FED technology is the lower power requirements of displays, i.e., 10 to 20 percent of AMLCD, due to the fact it is emissive and thus does not require a back light.

Notebook computers with FED screens could, when placed in large scale production, result in the ability to fly from the United States to Japan and not lose battery power of the device. This development can possibly accelerate the utilization of mobile computing, up to this point held in check because of cost and battery limitations.

2.   SOCIOTECHNICAL FACTOR

According to Porter (Porter et al. 1991, p. 20) society and technology are abstractions for a variety of entities with various levels of concreteness and aggregation. Porter describes a *sociotechnical system* as an open system which includes the following elements:

- technological devices and principles;
- scientific knowledge;
- institutions;
- individuals;
- financial resources;
- natural resources; and
- values.

Advances in microprocessors, network carrier lines, switching devices, data compression and decompression are the driving forces for many choices both technological and societal.   While microprocessors will benefit all distributive technologies, the key is receiving and network technologies. Microprocessors are changing televisions from dumb monitors to terminals on a distributive network. This advance, combined with advances in network technologies, i.e., high speed routers, switches and improved information protocols, will make televisions affordable interactive devices, delivering information and entertainment.

2.1   Sociotechnical System

Porter (Porter et al. 1991, p. 6) shows there exist relationships that influence the ability of the nation to absorb and utilize technology. This concept was further developed by Porter (Porter et al. 1991, p. 23) into a version of the sociotechnical system viewed as a delivery system. The boundaries of this technology delivery system are generally arbitrary. The sociotechnical change process of this system include:

- development of scientific knowledge;
- development of technological principles;
- development of prototypes;
- production and diffusion of technical devices;
- changes in characteristics and objectives of institutions;

- changes in characteristics and beliefs;
- changes in wealth and resources; and
- changes in values.

The development of scientific knowledge is not cyclic. In some instances an application or device may precede the actual scientific knowledge that underlies the device. Thus, developments can accelerate until a basic scientific barrier is reached. In some instances, it is this barrier which forces further scientific study, resulting in further developments.

Institutional change brought about by new technological developments does not necessarily maintain equilibrium. A case in point is the global impact of communications. The ability to rapidly exchange information between many individuals located in many diverse cultures has driven and will continue to drive, institutional change. Closed informational societies can lead to institutional instability.

Changes, in the state of resources, are not necessarily the result of conflict, but are, in an economic sense, driven by supply-demand considerations. One of the driving factors for precious metals, over the last decades, has been to use them in high technology industries. The use of these precious metals has not been as a hedge against financial system instability as much as the need for use in technology.

Sociotechnical change is more of a spiral process in which system changes adapt under causal influences (stochastic resonance) of its elements. One small change in a technological system can rapidly proliferate and have major societal changes. A case in point is the rapidity of the development of the Internet system from purely an exchange of scientific and technical knowledge to a potentially global paradigm shift due primarily to the development of the World Wide Web ("WWW") metaphor.

This is a non-linear system capable of chaotic behavior under certain initial conditions. A potential small scientific or technological development can cause the sociotechnical system to enter into a chaotic behavior. The advent of the development of the transistor by Bell Laboratories started a process that has been accelerating and changing many factors of everyday existence from communication systems to national security. Such developments and the cascading nature of the interactions with societal elements are not possible to predict over long periods. These non-linearities in the system, like any large scale system, have short predicable horizons, i.e., short term weather systems versus long term weather pattern prediction.

6

This sociotechnical system can take several states or outcomes:
- stabilized;
- incremental change; and
- discontinuous change.

It is the conflict between goals, values, and world views that may lead to incremental or disordered change.

2.1.1 System Elements

Every system can be viewed in a holistic manner by means of the elements composing it. Figure 1.1 shows a simple diagrammatic view of a sociotechnical system consisting of four basic elements: inputs; institutions and organizations; processes and outcomes. These elements interact to form the complex system in which technological developments are formed and utilized.

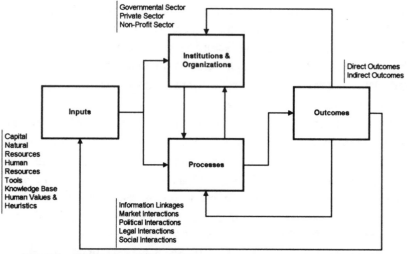

Fig. 1.1   Sociotechnical System View

Input

Inputs consist of capital, natural and human resources, tools, and a knowledge base that is derived from basic and applied research, and finally the system of human values and heuristics. In some instances, inputs are controlled by enterprises in which the technology is to be embedded. In other cases, stochastic resonance and cross impacts affect the inputs and make them difficult to control over the long term.

Institutions and Organizations

Institutions and organizations modify and control output from this system. These institutions are both public and private. Governmental agencies, through regulatory and fiscal actions, can bias technological developments and utilization. The thrust of governments to act as sensors of the Internet system, while based upon national cultural norms, may be difficult or impossible to implement due to the complexity of the underlying technological concepts.

Private sector enterprises either individually or in consortia (formal and informal) join to develop and implement the embedment of technological developments within the societal fabric. In some instances these organizations have also joined to prevent or constrain certain developments that may have been economically dysfunctional to their individual or joint interest.

Process

Institutions interact through various means that include: information linkages; market interactions; political interactions; legal interactions; and social interactions. These interactions, or processes, are on both a technical and human level. In some cases, these interactions may not be obvious as is the case where political lobbying by enterprises may result in substantial legislative changes which change the technological vector.

In a highly structured information society, data transferred between institutions can trigger actions that may have unexpected consequences. Such action can be demonstrated by the equity markets through *program trading*, a result of high speed computer analysis, that can cascade into very serious market sell-off. Similarly, chance meeting between several senior executives at a social function could lead to mergers, acquisitions and consortia resulting in major changes in the sociotechnical system output.

Outcome

The outcome of the sociotechnical system can have an effect on social and physical environments in which the technology is embedded. Consequences can be both intended (direct) or unintended (indirect).

The development of new technology is usually, but not always, undertaken to achieve a particular outcome. This intended outcome, such as the production of a new microprocessor, is the objective of the developmental process. Some developments, in the past, have been undertaken, not necessarily to achieve an obvious technological outcome, such as the development of the Strategic Defense Initiative ("SDI"), but to achieve a political end.

Throughout history, technological developments have resulted in unintended outcomes. The development of gunpowder by the Chinese for the joyous purpose of celebration resulted in weapons development which in many of the cities of the world still a thousand years later reaps social havoc. Similarly, the development of highly radioactive materials for military purposes has resulted in means to assist in alleviating human pain through the medical application of these elements.

2.1.2 Sociotechnological Mapping

The sociotechnological system proposed by Porter et al can be useful for conceptualizing the development of a single technology and the kinds of sociotechnical changes that could result. *Single* sociotechnical systems exist within a larger context of a technological industry, industries, nations and global constructs.

However, Porter's approach can be useful in providing a manager of technology with a means to understand the interactions of the various system components. This methodology can be useful in *mapping key players* who can impact the technological development.

Managers of technology rarely depict technological developments as vectors. All developments have both magnitude and direction at any particular instance. Depicting the essential *technological vectors* is an important element of managing technology.

Technological mapping should identify gaps resources, both physical and human. Many developments do not reach maturity due to the lack of a particular resource at a critical developmental stage of the technology.

Every system contains leverage points that can greatly magnify or diminish inputs. In the technological system, such leverage points, both physical and human, also exist. Key individuals or processes may hold the key to the next step in the development of a technology. It is important for the manager of technology to locate and emphasize these *technological leverage points*.

Technological enterprises do not develop in isolation. It is important for managers of technology to determine and highlight technological enterprises that can have impact on their technological developments. These technological enterprises can either be competitors or potential allies. Developments in other enterprises either similar or dissimilar can have major impacts. Many developers of software and hardware have been superseded by developments in small and large enterprises that they refused or did not recognize.

3.   TECHNOLOGICAL AND COMPETITIVE ADVANTAGE
Organizations and nations are faced with growing national and international competitiveness that places a strong emphasis on the following component measures:
- Productivity
- Rapid product introduction
- Quality
- Reliability

Technology is considered a measure of national power and an implement of public policy. In these terms *high technology* has been defined by Porter (Porter et al. 1991, p. 7) as:

"high-value-added products that tend to embody
the current state of the art and have a large
research and development – R&D – content."

The enterprise is the center of activity in asserting national competitiveness because, in our capitalistic democratic society, it is the firm in most instances that must deliver the technology. The recent changes in nations with planned societies have also shown that the enterprise is more effective in delivery of technology, products and services than the state.

However, the acceleration of technological change has placed stress on managers of technology due to:
- products that have decreasing market or operational life cycles (see Chapter Five - *Technological Life Cycles and Decision Making.*);
- increased concerns for safety, environmental effects and societal impacts;
- need to also deal with strategic and technical considerations simultaneously; and
- different investment decisions.

10

Management of technology for competitive advantage is directly linked to Research and Development ("R&D".) R&D must be directed toward technologies that have the highest probability to impact an enterprise's prime markets, i.e., therefore *true* identification of prime markets is crucial to any R&D program. The selection of R&D projects must understand customer needs and market dynamics. The resulting technologies must move through the total *developmental technological* cycle as beneficial innovations. Figure 1.2 shows the cycle for R&D. Technological change is one of the principal drives of competitiveness. This competitiveness permeates the total technological process (Hamel and Prahalad 1994, p. 212).

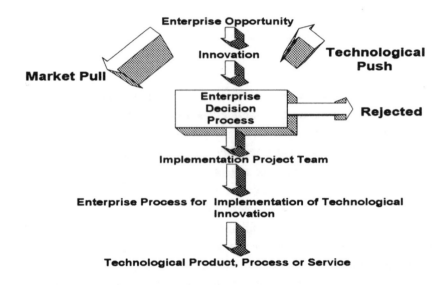

Fig. 1.2   Project-dominant model of the R&D process

Technological change plays a major role in structural change of existing industry, as well as in creation of new industries. It also serves to change the relationships in an industry and also among nations. Enterprises that have installed an effective system for encouraging and managing technology will be in a better position to develop and provide products and services (Edosomwan 1989).

A *social attractor* exists where many nations face a vicious circle of lack or underdevelopment of technology. This *social attractor*, can eventually result in a long lasting bifurcation among nations. Figure 1.3 shows the system that drives a *social attractor*. A similar *social attractor* exists within nations, even the United States, which is facing a bifurcation between a technically trained and untrained society.

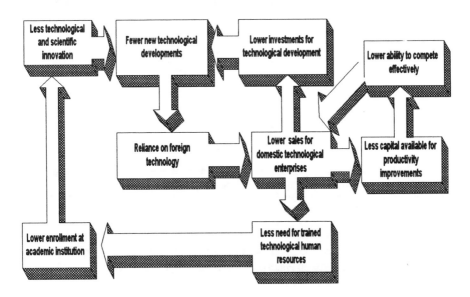

Fig. 1.3 Attractor model for technology and underdevelopment

4. LESSONS LEARNED

During the twentieth century, scientific and technological developments have progressed at an ever increasing rate. Lessons are learned from the progress and pitfalls that have been experienced. Lessons learned can be divided into scientific and technological developments. Figure 1.4 shows the relationship according to Betz (Betz 1987) between these two areas.

4.1 Scientific Developments

Scientific developments appear to follow various patterns which proceed from early discoveries to eventual utilization of the concepts within a technology. Rarely do scientific developments arise from an intended technological application. In fact, scientific discovery's basic intention is to understand physical phenomena.

12

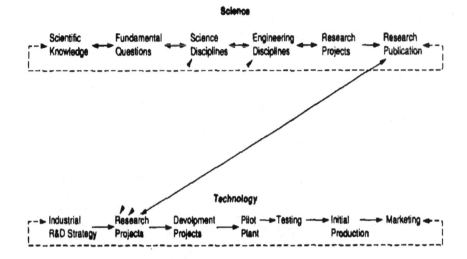

Fig. 1.4   Relationship between science and technology (*Source:* Betz 1987, © 1987, Prentice Hall Law and Business, Assigned to Aspen Law & Business, A Division of Aspen Publishers, Inc. Reprinted by permission.)

4.1.1 Nuclear Physics

It appears that a chain of scientific events in various locations and not developed under a single sponsorship is the form of this type of scientific development. The chain can be described as a progression of steps, sometimes rapid, sometimes halting. It does appear to require the accumulation of sufficient knowledge that eventually leads to a rapid cascade of events and discoveries. The steps include: conceptualization based upon prior knowledge and techniques; experimentation; which leads in many instances to the need for a new form of instrumentation, followed by additional experimentation and additional conceptualization that cascades to a finalized concept that can lead to technological applications.

Table 1.1 shows the major events in the development of nuclear physics during the twentieth century. The first event, i.e., the 1905 seminal paper by Einstein that presented the basis of understanding of the basic principles of nuclear energy, was a scientific breakthrough and was totally unpredictable. This purely analytical paper *triggered* the growth of nuclear physics in this century. However, it was based upon developments in mathematics and physics in the ninetieth century which Einstein brought together in a new theory of the equivalence of mass and energy, a theory that was a violation of acceptable scientific theory. Einstein *took a step* into the unknown but based upon a sound mathematical basis.

Table 1.1
Major Events in Nuclear Development

Year	Event
1905	Mass-energy equivalence. Publication of a paper by Einstein established the equivalence of mass and energy
1906	Isotopes of radioactive elements. Discovery of chemically identical elements with different radioactive properties.
1911	Atomic structure. Experiments by Rutherford showed that the mass of an atom is concentrated in a positively charged nucleus.
1913	Isotopes of nonradioactive elements. Discovery of isotopes through differences in physical properties.
1919	Ejection of protons from nitrogen. First artificially induced nuclear reaction.
1919	Mass spectroscopy. Accurate determination of the masses of isotopes
1920s	Mass defect (packing fraction). Discovery that the mass of a nucleus is less than the sum of the masses of the constituent particles.
1932	Discovery of the neutron. New particle, same mass as the proton, but sharing no electric charge.
1938	Fission of uranium nucleus. Uranium atoms split into roughly equal halves.
1939	Chain reaction hypothesized. If neutrons are emitted during fission, further fission can take place.
1942	Chain reaction produced, Actual demonstration of fission by neutrons emitted from earlier fission.
1945	Atomic bombs. First use in warfare.
1956	Commercial nuclear power generation. Actual power plant generating electricity from nuclear energy.

(*Source:* Martino 1993, p. 189. Reprinted by permission of The McGraw-Hill Companies.)

The discovery that not all chemically identical elements were physically similar by Boltwood in 1906 and McCoy and Ross in 1907 showed that the radioactive elements ionium and radiothorium where chemically identical with the element thorium. Although they had different masses, these similar elements were termed *isotopes*[1].

---

[1] The definition of an isotope according to the *Concise Oxford Dictionary* is: "*one of two or more forms of an element differing from each other in relative atomic mass, and in nuclear but not chemical properties*"

14

The scientific developments progressed in 1919 with the first artificially induced nuclear reaction. However, the source of this energy required a new instrument, i.e., a mass spectrometer, which allowed researchers to determine the masses of atoms. In 1919, Aston's[2,3] mass spectrograph made it possible to distinguish between isotopes and to measure the different atomic weights of these same elements. After this, the developments came more rapidly, especially with the development of electrical machines to accelerate particles to speeds that would split atoms so as to cause the artificial transmutation of elements, i.e., the true *philosopher's stone*. It is important to note that these machines worked on the same principles as the mass spectrograph.

The electrostatic generator of Cockcroft[4] and Walton was one of the first scientific devices in the chain of events in the development of nuclear physics. In 1930 Lawrence[5] invented the cyclotron that accelerated protons along a circular and a far longer path than the linear accelerator developed by Cockcroft and Walton. These developments led to Chadwick's[6] discovery of the neutron in 1932. Throughout the 1930s the world-wide scientific community conducted numerous experiments in which they bombarded various elements with neutrons. In 1938, Hahn[7] and Strassman, two German physical chemists, discovered they could split the uranium atom into smaller fragments that gave off considerable energy. The element uranium had been identified and named over two hundred years before and isolated fifty years

[2] **Aston Francis William** (1877-1945): British chemist and physicist. In 1922 he was awarded the Nobel Prize for developments that led to the discovery of isotopes in nonradioactive elements.

[3] All biographical material has been abstracted from Microsoft's® *Bookshelf '95*.

[4] **Cockcroft, Sir John Douglas** (1897–1967): British physicist. He served as fellow and professor of natural philosophy at Cambridge, he also directed (1946–59) the British Atomic Energy Research Establishment at Harwell. In 1952 he shared with **Ernest Walton** the Nobel Prize in physics for their pioneer work in transmuting atomic nuclei by bombarding elements with artificially accelerated atomic particles.

[5] **Lawrence, Ernest Orlando** (1901-1958): American physicist. In 1939, Lawrence won the Nobel Prize for the development of the cyclotron.

[6] **Chadwick, Sir James** (1891–1974): British physicist. He worked on radioactivity under Ernest Rutherford and was assistant director (1923–35) of radioactive research at the Cavendish Laboratory, Cambridge. For his discovery of the neutron he received the 1935 Nobel Prize in physics.

[7] **Hahn, Otto** (1879–1968): German chemist and physicist. Noted for important work on radioactivity, he received the 1944 Nobel Prize in chemistry for splitting (1939) the uranium atom and discovering the possibility of chain reactions.

later. Two refugee Austrian scientists working in Copenhagen showed, analytically, that Hahn and Strassman's experiment with the splitting of the uranium nucleus, when struck by a neutron, splits into two nuclei of approximately equal weights, one of which was an unstable barium nucleus; at the same time a considerable amount of energy is released.

The next steps quickly led Bohr[8], Fermi[9] and others to the concept of a chain reaction -- the rest is technological application leading to the first atomic weapons and nuclear energy production. In most instances the work of these talented scientists was recognized through the award of a Nobel Prize.

4.1.2 Quantum Physics and Relativity
There exists an event chain for the developments arising from quantum physics and relativity. Initial experimentation which did not confirm accepted theories was the first link in this chain. This was followed by development of supporting analytical theories that were independent of experimentation. These analytical theories and data then were incorporated into the academic knowledge base which was the basis for the training of scientists and engineers knowledgeable about scientific developments that initially appear to have little practical application. The eventual employment of the knowledge base to solve technological application problems which in turn led to further developments is the final link in the event chain.

The development of quantum physics and Einstein's theories of relativity led to technological applications far from the concepts envisioned by the theoreticians. In 1901, Planck[10] presented his quantum mechanical theory, which was developed to reconcile the observed distribution of energy in the spectrum of heat radiated from a hot *black body* problem with accepted theory. Planck found a formula that fitted experimental results. According to his quantum theory, radiant energy can be emitted or absorbed only by discrete units. Planck did not really think much of his theory. To Planck, it was a mere temporary expedient. Einstein's theories rejected Newton's basic postulate of the absoluteness of space and time and of the axiom of

---

[8] **Bohr, Niels Henrik David** (1885-1962): Danish physicist. In 1922, he won the Nobel Prize for investigating atomic structure and radiation.
[9] **Fermi, Enrico** (1901-1954): Italian-born American physicist. In 1938, Fermi won the Nobel Prize for his work on artificial radioactivity caused by neutron bombardment. In 1942 he produced the first controlled nuclear chain reaction, in a squash court at the University of Chicago.

conservation of mass. Quantum theory and relativity found no practical application for many years. However, analytical and experimental work continued in institutes and universities. During the same period as these scientific developments were occurring, the radio industry and its supporting electronic technology were undergoing rapid growth.

To provide necessary research and development for the growth of these industries, technologists were recruited from the newly trained physicists and engineers who were familiar with the work derived from Planck's and Einstein's theories. The training of these engineers and physicists led to a number of technological applications that would not have been possible without the basic research in physics. These developments include:

- radar
- television
- transistors
- lasers

Each of these technologies was based upon the basic theories of quantum physics and they also led to further cascading of technological developments. This cascading has continued throughout most of the twentieth century and will likely have major impacts in the next century.

4.1.3 Computer Science

In the case of computer science, we can postulate a different event chain scenario for the developments. This event chain appears to have been initiated by the conceptualization of a scientific device, but the supporting theory and technology did not exist or was in an embryonic form. The gap remained in scientific development as supporting theory and technology proceeded along independent paths. However, major societal problems, i.e., war, required a device to meet particular needs, and finally scientific and other technical personnel using the background and scientific knowledge bases incorporated these concepts to formulate a dual use device.

Today, enterprises increasingly rely on information technology to perform their day-to-day operations, and as a source of new products and services. However, the genesis is based upon developments in computer science.

---

[10]**Planck, Max Karl Ernst Ludwig** (1858-1947): German physicist. In 1918, Planck won a Nobel Prize for discoveries in connection with quantum theory.

According to Cardwell (Cardwell 1994, p. 467-484), the scientific motives were the solution of specific scientific problems, to build advanced devices, and to *push* against the frontiers of knowledge. In the nineteenth century, Babbage[11] proposed a machine specifically for scientific purposes. He proposed, but did not build, a scientific computer that would deal with negative and positive numbers and complex mathematical functions.

However, nothing happened until the 1930s when Konrad Zuse in Germany, with his own resources, started to build a mechanical computer. In 1937, Aitken published a specification for a computer machine together with a survey of previous work in the field. During the same period Turing published his *On Computable Numbers*, a classic in its field that also showed that there were unsolvable problems. Turing conceived the abstract concept of a *universal computer*. The beginning of World War II accelerated the developments of computer science.

- In Germany, Zuse, who was supported by the German government, built an electromechanical computer to determine ballistic calculations.

- In England, Turing[12] and Von Neumann [13] working on the solution of the *Enigma code* problem built an early electronic machine based upon the binary concept of Von Neumann.

- In the United States, the University of Pennsylvania in co-operation with the U.S. Army was working on an analog computer for ballistic tables. A member of the staff, Mauchly, thought of replacing the mechanical analog computer with a digital electronic computer. Mauchly and Eckert (also a member

---

[11]**Babbage, Charles** (1792-1871): British mathematician and inventor of an analytical machine. Babbage's fame rests on his attempts to develop a mechanical computational aid he called the "analytical engine." Although it was never constructed and was decimal rather than binary in conception, it anticipated the modern digital computer. He was a scientist with extremely broad interests.

[12]**Turing, Alan Mathison** (1912–54): British mathematician and computer theorist. Turing's early work in predicate logic led to a proof (1937) that some mathematical problems are not susceptible to solution by automated computation.

[13]**Von Neumann, John** (1903-1957): Hungarian-born American mathematician noted for his contributions to game theory and quantum theory.

of the staff) built the first electronic computer, Electronic Numerical Integrator and Calculator ("ENIAC") in 1943. Thus, in this instance, the demarcation between scientific and technological application is not as distinct as in other scientific developments.

4.1.4 Material Science

Material science also demonstrates the flow of an events chain leading to technological products. This class of scientific development seems to follow the scenario chain which starts with experimental work that demonstrates a phenomenon that has no associated theory and that would be difficult to implement in the form of a technological application. Eventually a workable theory is developed to explain this phenomenon. New materials or devices are then discovered that increase the capability of the phenomenon, but still the usability, capability, reliability and economics are not yet demonstrated to have widespread commercial use. Finally, the technologist either must wait for better capabilities or develop ways to circumvent particular technological roadblocks, like better manufacturing techniques.

One of the most important breakthroughs in physics has had profound impact on a particular branch of material science, i.e., superconductors. In 1911, the Dutch physicist, Onnes, discovered that electrical resistance in mercury vanishes when it is cooled to temperatures close to absolute zero (-460°F or 0K). Onnes termed this phenomenon superconductivity. Other experimenters soon discovered that other metals and alloys become superconducting at very low temperatures. The temperature at which a metal becomes superconducting is called the critical temperature. The higher the critical temperature of a metal or alloy, the more easily it can be used in technical applications.

In 1957, forty-six years after the discovery of superconductivity, Bardeen[14], Cooper and Schrieffer proposed a theory to explain the basic phenomenon for which they were awarded the Nobel Prize. However, it was not until 1987 that two IBM researchers, Müller[15] and Bednorz[16] developed

---

[14]**Bardeen, John** (1908-1991): American physicist. Bardeen shared a Nobel Prize with **Leon N. Cooper** (Born 1930) and **John Robert Schrieffer** (Born 1931) in 1956 for the development of the electronic transistor and in 1972 for a theory of superconductivity.

[15]**Müller, Karl Alex** (Born 1927): Swiss physicist. Müller shared a 1987 Nobel Prize with Bednorz for pioneering research in superconductivity.

materials that become superconducting at substantially higher temperatures than liquid helium. The new ceramic materials become superconducting between 90°K and 120°K, i.e., above the boiling point of liquid nitrogen, which is inexpensive and easy to maintain. However, in 1988, when this discovery was made, no satisfactory theory existed to explain the mechanism of superconductivity in ceramic materials.

The future of widespread use of superconductors still awaits the development of technologically usable materials that are reliable and cost effective.

4.2   Technological Developments

Unlike scientific developments, technological developments are a problem oriented activity. Scientific developments are concerned with increasing the knowledge base about a particular phenomenon. Technological developments are concerned with the solution to a particular problem or creation of an application to perform a particular function. However, not all answers or developments result in useful or acceptable solutions that move on to Martino's next level of development, i.e., widespread use.

According to Edosomwan, technology leadership is exerted through *function* in producing products with advanced performance or features (Edosomwan 1989, p. 143). If commercial benefits of technology innovation are to be retained by the enterprise, the organization must exert both technology and market leadership.

4.2.1 Semiconductor Technology

The semiconductor industry has over the past several decades experienced phenomenal growth. This growth has been characterized by a cycle of domestic and international competition. During the early 1990s, the semiconductor industries had substantial growth in profit margins. Internationally, Taiwan's semiconductor industry profit margins grew from 16 percent in 1990 to nearly 60 percent by 1995 (Economist 1996). However, the semiconductor industry has become exceedingly capital intensive with manufacturing facilities, also known as *fabs*, costing one to three billion dollars per facility. This capital intensiveness has served to bring

---

[16]**Bednorz, J. Georg** (Born 1950): German physicist. He shared a 1987 Nobel Prize for pioneering research in superconductivity.

together several semiconductor manufacturers to form consortia to design and fabricate new semiconductor designs. These consortia at times achieved their objectives and in other instances failed. The PowerPC and SEMATECH consortia are examples of these unions within the semiconductor industry.

The development of the PowerPC by the consortium of IBM, Motorola and Apple to counter Intel's dominance on the microprocessor market is an example of competitive business arrangements based on technological developments. Similarly SEMATECH was formed to counter the dominance of the Japanese microcomputer industry. However, while SEMATECH consortia resulted in a major change in the dominance of the international microcomputer industry, the PowerPC development has seemed to flounder; the question we must ask is:

> Why does one form of technological development succeed in one case and basically is failing in another case?

Power PC Lesson

The PowerPC project grew out of a desire on the part of IBM, Motorola, and Apple to create a family of microprocessors with the potential for gaining a significant market share and capable of supporting aggressive development in the future (Lerner 1994), and through technology gain a competitive advantage over a major market leader. The consortium hoped that a new direction in microprocessor architecture would not only cut into Intel's market share, but change the nature of computing. The three members of the alliance agreed that the new architecture would be based on IBM's existing POWER ("Performance Optimization With Enhanced RISC") microprocessor series, modifying it, but insuring backward compatibility for application programs. The agreement on maintaining backward compatibility with the POWER architecture resulted in *an evolution, not a revolution*, in processor design.

Whatever the architecture, a new chip has to have circuits that are faster, and smaller. The speed of microprocessors has been increased through the introduction of dynamic complementary metal oxide semiconductor ("CMOS") circuits. As the size of the circuits decreases, changes in physical parameters create design challenges. Electrical noise is another problem that is exacerbated by the use of smaller and faster circuits. The problem of electrical noise is more prominent as succeeding chips in the PowerPC series aim for increasingly faster clock frequencies. In this case, three enterprises, IBM, Motorola, and Apple, each with different cultures and management styles have worked on these and other development problems as members of the consortium.

For each POWER instruction, there was a choice: either to hardwire it into the chip, or emulate it in software. In the end, the controversy came down to deciding which choice would be the most efficient or least painful. A more serious problem was that, while incompatibility between the instruction sets of the POWER and PowerPC architecture would be eliminated when programs were recompiled, that was not possible with vendor-supplied legacy[17] programs.

Furthermore, there were no simple and objective ways to make decisions, a common situation in both software and hardware design. All members of the consortia were aware that any change would incur costs to IBM in compatibility fixes, whereas Motorola and Apple were moving to an entirely new architecture, and less concerned with backward compatibility. According to Lerner, the debates were driven by professional pride in what had been done in the past. The technical challenges in the PowerPC development were not in small but controversial changes to instructions sets, but rather in the move to the 64-bit architecture, a major change. The second major change was the introduction of multiprocessing capabilities.

IBM, in September 1993, announced the RISC System 6000 model 250 series of workstations featuring the PowerPC 601 processor. In March 1994, Apple announced its line of PowerPC based Power Macintosh. However, by August 1995, the industry magazine, *Information Week* (Gillooly 1995) headlined an article entitled "PowerPC Shorts Out - New chip fails to spark large sales, so vendors target core markets."

The PowerPC had been designed to help Apple enter the enterprise market and spearhead IBM's desktop reassurance. Instead, sales for their computers based on the new chip were not meeting even conservative sales expectations; according to an industry executive, the "*Sales have been pretty abysmal.*" (Gillooly 1995)

---

[17] Legacy programs are computer programs which were optimized for prior chip architectures.

SEMATECH Lesson

A successful use of consortia to convert technological advancement into a competitive advantage is illustrated by the Semiconductor Manufacturing Technology, Inc. ("SEMATECH") that was established in 1987 to reverse the loss to Japan of semiconductor manufacturing leadership. SEMATECH was a unique experiment in American industry-government co-operation. The enterprise was given the national mission of quickly restoring, to the United States, world leadership in semiconductor manufacturing.

SEMATECH was incorporated with fourteen high technology companies representing eighty percent of the national capacity for semiconductor manufacturing. The actual goal of SEMATECH was to improve the state of U.S. semiconductor manufacturing technology, especially improving the current generation of equipment and the equipment that would come into widespread use within five to eight years (Gover 1993). The emphasis of SEMATECH has been in manufacturing technology, and not the products made with that equipment. SEMATECH is a horizontal consortium focused on strengthening upstream suppliers for its member enterprises. This is relatively easy to achieve because SEMATECH's focus is not on its members' products, but on its suppliers of manufacturing equipment. SEMATECH has indeed achieved its goal of reversing manufacturing semiconductor leadership despite some setbacks (Grindley et al 1994).

SEMATECH has demonstrated that government-sponsored R&D programs can have economic impact on the competitiveness of a critical industrial sector provided the structure is well designed (see Chapter Six - *Enterprise Structure*). It is also necessary that there exists constructive co-operation among members of the consortia without inhibiting competition among its members by working on problems upstream from the members' markets. SEMATECH also illustrated that linking semiconductor manufacturers and semiconductor-manufacturing equipment makers accomplished the positive attributes of Japan's keiretsu without the negatives.

The question we must ask is why SEMATECH
succeeds and PowerPC seems to be failing?

Possible Answer

Grover has synthesized a competitiveness-commercialization model. Figure 1.5, may give us the answer to the question of why one consortium succeeds and another fails (Gover 1993). The dominant themes from the SEMATECH lessons include that customer need and market forces must influence the direction of any technological development. The success also illustrated that co-operative working teams spanning research to manufacturing, preferably working together in a shared facility, are more likely to succeed than separate teams in separate facilities. It also demonstrated that communications must be facilitated by all practical means. Non-duplication of research was an important element in maximizing the use of scarce resources. It also illustrated that emphasis on today's problem product and process improvement drives commercialization against seeking major breakthroughs.

The PowerPC consortium did not focus on all the elements necessary for maximum competitiveness in high technology industries. The consortium did not strive for all four elements of the competitiveness model illustrated by Figure 1.5 (Gover 1993).

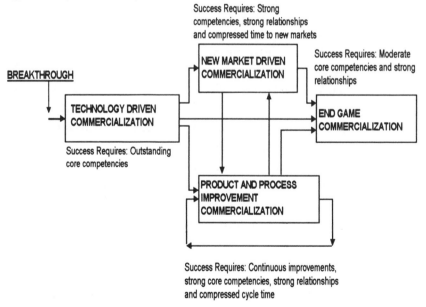

Fig. 1.5   Commercialization model for technology (Source: Gover 1993, © 1993 IEEE. Reprinted by permission.)

4.2.2 Information Storage Technology

The impact of technological advancement on competitive advantages is also illustrated by the growth in information storage technology. A review of this technology industry indicates that it can be summarized as (Jabbar 1995):

- technology-driven due to rapidly changing products and processes;
- market-driven due to limited market size and product selection;
- quality-driven due to OEM customers and quality of vendors for components;
- competition-driven due to the need for relentless fight for market share; and
- entrepreneur-driven since the companies are controlled by principals with individual enterprise vision.

Information storage technology has moved from the early 1940s' low density storage systems to the very high density devices in 1995. The early computers used magnetic tape which been developed in Germany during World War II, to replace magnetic drums for back-up storage. As computers became more complex and faster, magnetic tape was not sufficient for rapid storage and retrieval. This complexity coupled with speed led to the introduction of the Winchester disk drive technology in 1956. Disk storage capacity has grown significantly since 1953. In 1953 IBM magnetic storage tape stored 1.4 kilo-bits per square inch, by 1981 this had increased for disk drives to 12,000 kilo-bits per square inch (Freeman 1982).

In 1982, IBM introduced the Personal Computer ("PC") which started another cycle of growth. With rapid progress in computer technologies, especially in PCs, the need for mass data storage devices rapidly multiplied. Since 1956, when the first disk drive was developed by IBM, disk drive technology has gone through a series of rapid transformations in form-factor resulting in various smaller sizes of drives. The capacities and performance of disk drives doubled almost every two to three years during the initial developmental period of this technology. Recording densities have also gone up by more than 1,000 times and data access performance has gone up by at least 15 times over the last two decades (Jabbar 1995). What made this progress possible is a combination of developments in:

- materials for heads and media;
- electronic for data communications;
- head-positioning systems;

- innovative designs for spindle motors; and
- new manufacturing and process engineering technologies.

The disk drive industry is both technology and market driven. The upstream markets for the disk drive industry consist of markets for personal computers ("PCs"), software development and microprocessors that are important to disk drive technology and markets.

To fully understand the technology issues of disk drives, the following technological developments are important:

- heads and slider development;
- media and substrates;
- actuator and head suspension;
- spindle motor and its control;
- read and write electronics and data channels; and
- computer interfaces.

Figure 1.6 shows the various components of a modern PC disk drive. A number competing technologies exists that can seriously have an impact on the disk drive technology industry including:

- massive storage capacities of new tape drives;
- massive storage capacities of optical drives; and
- flash memory cards with tremendous access speed.

There are a number of characteristics of the disk drive technology industry:

- Short product life cycle that has been averaging in the 1990s 18 months or less. This requires expensing associated cost for tooling, etc. during a shorter life cycle.
- Low automation in the disk drive industry that results in high production cost.
- Vendor concentration, i.e., small number of vendors for platter medium and read-write heads of the disk drives. This tends to cause increased cost of inputs.
- Disk drives are primarily sold to original equipment manufacturers ("OEM"). Thus the markets for the disk drive manufacturers tend to be very limited, resulting in high opportunity costs.
- Most disk drive manufacturers have fragmented management that results in high communication cost.

26

- Lack of process technology due primarily to rapidly changing technology results in lower yields and higher scrap cost.

The computer industry and associated technologies have moved through Martino's development phases to become widespread, and have moreover become a technology industry that has turned its basic output products into commodities. This development of commoditization, for a technology industry, results in continuous reduction of profit margins and at the same time fuels innovations in technology. The computer industry and associated technologies have been globalized with facilities, markets, vendors and customers located around the world and with the complexity that is engendered (Jabbar 1995). This globalization results in problems of management and international technology trade.

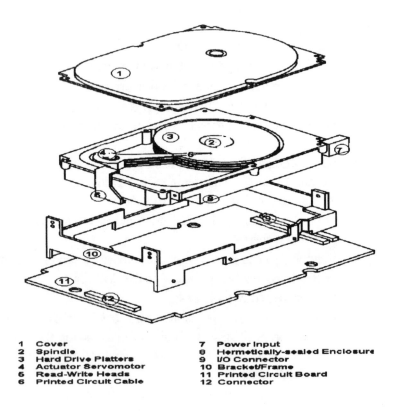

1	Cover	7	Power Input
2	Spindle	8	Hermetically-sealed Enclosure
3	Hard Drive Platters	9	I/O Connector
4	Actuator Servomotor	10	Bracket/Frame
5	Read-Write Heads	11	Printed Circuit Board
6	Printed Circuit Cable	12	Connector

Fig. 1.6   Components of a modern PC computer disk drive

Technology is embedded within a societal and competitive framework. The management of technology is similarly embedded and affected by these environments. It is important to understand the process within these environments: a *systems approach* conceptualization is important. The concept on a basic level is the application of the *systems approach* to management and is emphasized throughout academic programs of studies.

5. ENVIRONMENT

5.1 Stages of Technology
Technology does not go directly from its creator to immediate application. It passes through a number of stages that represent greater degrees of practicality or usefulness. The following is a general set of stages in this creation to application process (Bright 1968; Martino 1993, p. 9):
- scientific findings;
- laboratory feasibility;
- operating prototype;
- commercial introduction or operational use;
- widespread adoption;
- diffusion to other areas; and
- social and economic impact.

5.2 Scientific Findings
Scientific findings represent a *knowledge base* from which solutions to specific problems can be achieved. In this initial technology stage, the pre-technology exists as an understanding of some phenomenon, the properties of some material, the behavior of some force or substance, or some other scientific characteristic. At this stage, this knowledge is not capable of being utilized to solve a problem or carry out a function.

5.3 Laboratory Feasibility
At this point in the process, the scientific finding has been identified so it can be applied to the solution of a specific problem, and a laboratory model has been developed. Clearly no natural or physical laws are violated by the application, and it is capable of performing the desired function or solving the problem of concern, but only under laboratory conditions. The application at

this stage may be described as a *breadboard model*. At this stage, the technology would not be used in a commercial or operational sense, i.e., apart from the test environment.

5.4 Operating Prototype

During the prototype stage it must be capable of operating within the intended environment by the technology's intended users, i.e., it must function satisfactorily. Software developers usually assign the term *Beta version* for the initial trial release of a technological development. Prior to the final release on Windows 95 in August 1995, Microsoft® had over 200,000+ *beta* testers.

5.5 Commercial or Operational

The next stage in the technology cycle is the use in a commercial or operational environment. In a commercial environment the technology faces its *true* economic test, i.e., economic feasibility. This is a crucial period for any technology or product, i.e., *Edsel Test*. The "first or Version 1.0 production model" is often considered the point when the technology or product has reached this stage.

Non-commercial technology, such as that developed and implemented by governmental agencies, is represented at this stage by it "first use" in its intended operational environment.

5.6 Widespread Adoption

After a technology has demonstrated that it is technically and economically superior to whatever else was used in the past, i.e., automobile versus a horse drawn cart it then replaces the prior technology rapidly. This stage varies with each technology and some technologies never reach this stage and maintain only a *niche* market position. An example is the use of xerography which rapidly replaced all other forms of copying written and graphic material.

5.7 Diffusion to Other Areas

Diffusion of a technology happens when it has not only dominated the application area in which it was first adopted but has been adopted in other areas. A case in point is the first adoption of jet propulsion to military aircraft, then the use by the commercial airline industry. At this stage, the technology has been adopted for purposes to which the earlier technology was never applied; that is, the technology supplanted some earlier technology, such as the replacement by transistor for vacuum tubes and now widely used in appliances, vehicles, cameras and other applications.

5.8 Social and Economic Impact

When a technology reaches the social and economic impact level, it has changed the *world view*. Television, Internet and other communication systems have caused major social changes. Not all technologies reach this stage directly. However, each technology has both direct and indirect impacts which taken together can cause *cascaded events* or world view impacts. These linkages can start with a minor scientific discovery which over time changes world views and from which arises new industries, concepts and other unforeseen developments.

6. SYSTEMS APPROACH TO MANAGING TECHNOLOGY

The systems approach represents a new approach to managing technology. The *new* system or holistic way looks at the world in terms of sets of integrated relationships. The systems view gives us a way of looking at complex management problems. It is a mode of organizing existing findings concerning the concept of systems, and systemic properties and relationships.

Successful managers of complex systems should possess the following qualities:

- understanding of the technology of their *business*, that is, it is important that managers have or be trained in the technologies which they must manage;
- understanding of the *basic concepts of management*, in many instances not having a firm grounding in these concepts will likely result in reduced technological performance;
- interpersonal style that facilitates their ability to get things done through others; intelligent delegation is the essence of many successful technology managers; and
- abilities to conceptualize and to operate using the systems approach, that is to take a broader or holistic view.

Most realistic management problems involve systems and the way they change. The *systems* concept has had a substantial impact on both the planning and the implementing functions of management. A *system* is defined by the *Concise Oxford Dictionary* as:

"complex whole; a set of connected things or parts; an organized body of material or immaterial things."

In the early 1960s researchers developed ideas on what eventually became known as *system analysis*. This methodology viewed that problems needed to be seen in relationship to the underlying systems.

System analysis has evolved from the early methodology:

1960s Systems Methodology for Problem Solving

- Define objectives
- Design solutions to meet objectives
- Evaluate cost-effectiveness of solutions
- Decide on the best solution
- Communicate the system solution
- Establish performance standards

1970s -- 80s Systems Methodology for Problem Solving

- Formulate the problem
- Gather and evaluate information
- Develop potential solutions
- Evaluate workable solutions
- Decide on the best solution
- Communicate the system solution
- Implement the solution
- Establish performance standards

The value of the systems concept to management of technology can be seen in terms of the elements of the manager's job of overall effectiveness and conflicting organizational objectives.

The systems approach to management of technology decision making requires the use of objective analysis of decision problems. The systems approach to technology management problem solving requires having the perspective to deduce important system variables and the relationship between them.

A general statement of this approach is given by the following phases (Flood and Jackson 1991, p. 50):

- Creativity

- Choice
- Implementation

6.1 Creativity

There are various ways to assist technology managers creatively about enterprises. An enterprise can be viewed as:

- closed system such as a *machine;*
- open system such as an *organism;*
- learning system such as a *brain* or *neural net;*
- emphasis in norms and values such as a *culture;*
- unitary political system such as a *team;*
- pluralist political system such as a *coalition;* and
- coercive political system such as a *prison.*

The view of the organization has an impact on its approach to managing technology. A combination of approaches or views, i.e., an organization that is open, learning and unitary in terms of application will likely result in maximizing the creativity phase.

6.2 Choice

The task during the *choice* phase is to choose an appropriate system-based technology or technologies to suit particular characteristics of the problem revealed by the examination conducted in the creativity phase. The most probable outcome of the *choice* phase is that there will be a *dominant* technology chosen to be modified or developed further in use as more information from the environment is obtained.

6.3 Implementation

The *implementation* phase is responsible for taking the chosen technology and *brings it to market* or utilization. During this general phase a number of actions are required by the organization. It is at this time that disconnect can occur between the original vision of the development and its translation to a market acceptable process. The Figure 1.7 maps the process of creation to application to the general system phases.

32

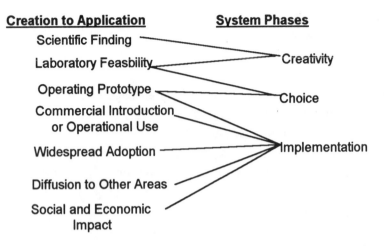

Fig. 1.7 Mapping of creation to application processes to system phases

7. PRECURSORS OF TECHNOLOGICAL CHANGE

There are various precursors of technological change (Martino 1993, p. 201).

- Incomplete inventions: These may demonstrate the feasibility of some technology but may require other elements before they can be deployed economically or in a particular environment.

  Example: Controlled fusion

- Development of performance-improving supporting technologies needed by the basic technology.

  Example: Lasers for high capacity fiberoptics for wide band communications

- Emergence of cost-reducing technologies needed by the basic technology

  Example: High volume and or high quality integrated circuit manufacturing techniques which were needed to make low cost personal computers possible

- Availability of complementary technologies: Complementary technologies are those that the basic technology must interact with to be useful.

  Example: Satellite position systems used to develop electronic highway control.

- Prestige or high performance application: Use of a technological advance in a prestige or high performance application before transition into general use provides the exposure to larger market segments.

Example: Use of technology in race cars before introduction into automobile production

- Incentivization: The requirement of an incentive for using the technology such as reduced cost or elimination of externalities can also be viewed as a precursor. This is the approach being used by a number of governments to utilize technologies that may at early market stages not have sufficient basic economic incentives, but may have other societal benefits that they believe should be encouraged.

Example: Passage of the Clean Air Act in the United States leading to the development of clean coal technologies

8. PRIVATE SECTOR

In a societal sense, a business enterprise is expendable. Customer satisfaction usually determines enterprise survivability. This is in contrast to a governmental or non-profit organization, such as a hospital or university, which usually remains in business no matter how poorly they satisfy their customers.

If a company wants to stay in business, it must continue to satisfy its customers. It must anticipate changes in the customer's wants and needs, as well as changes in the ways the wants and needs can be satisfied. A new company must satisfy its *stakeholders*, i.e., bankers, investors, and principal staff who have invested their time, creativity and energies in an unproved enterprise.

An important consideration in the success of a business is to accurately identify the nature of the business enterprise. This is usually done in only superficial terms that often are too narrow, leaving the enterprise vulnerable to technological shifts and changes in customer needs.

The nature of an enterprise's business can be obtained by observing:

- Specific functions performed by the enterprise in a market and the objective of these functions in terms of the enterprise's objectives such as banking.
- Specific product or service provided to clients or customers as the case of microprocessors.
- Specific process utilized in the case of Internet providers

- Specific distribution system utilized such as employed by mail order companies
- Specific set of skills as used by a consulting company
- Specific resource utilization as when a paper manufacturer uses wood pulp and recycled paper

Once the type of business is determined it is possible to evaluate the consequences of technological change. Some technological changes can lead to the disillusionment of the enterprise. Therefore, it is important that managers be very alert to technological changes that can alter the way they do business. These changes, if not anticipated, may leave the enterprise unable to compete with companies in the same business. An example is Wang Computer and Digital Equipment Corporation -- Wang went into bankruptcy and Digital Equipment Corporation lost significant market share because they both failed to adequately see the changes by the small personal computer on corporate buyers based upon the Intel chip technology.

Technological change can alter the fundamental nature of an enterprise. These impacts can occur in a number of areas (Martino 1993, p. 270).

Function: An enterprise is vulnerable either to a technological change that makes the function unnecessary or to one that performs the function in some other manner.

Examples:

- Transportation: roll-on-roll-off trucking
- Utilities: electric versus gas lights
- Shipping: container shipping

Product: A technological change may allow a different product to be used to replace another product to accomplish the same or an improved function.

Examples:

- Transistor versus vacuum tubes
- Automobile versus horse drawn carriages
- CD ROM versus printed material

Process: A new process for manufacturing a product providing some service that can almost directly replace an older process.

Examples:

- Federal Express versus U.S. Postal Service
- Continuous casting of utility wire versus welding sections
- Bioprocessing versus chemical processing

- Direct coal injection for steel manufacturing versus the use of coke

Distribution: Technological change has altered the way many firms distribute their products.

Examples:
- Airfreight versus trucking and ocean shipping
- Internet versus printed media

Skills: The skills mix in an enterprise may drastically alter with technological change. This can occur because of new processes or new products.

Examples:
- Computer controlled manufacturing versus manual control
- Computer typesetting versus individual typesetter

Raw Materials: New technology can result in a change in material used for both process and products. Man-made materials in many instances have replaced natural materials.

Examples:
- Reinforced plastic in cars versus metals
- Composites versus metals

Management: The manner in which a firm is managed can also be affected due to technological changes.

Examples:
- Groupware versus individual products
- Virtual organization versus physical organization
- Management sciences versus solely managerial judgment

Support: Operational activities such as hiring, payroll and other support functions can be impacted by technological change.

Examples:
- Behavioral technology versus standard hiring practices
- Management Information Systems versus manual payroll and record keeping.
- FAX versus phone support

However, technological change can have an impact on one or more of these enterprise aspects including its managerial and support activities.

9. GOVERNMENTAL SECTOR

The structure of society requires governmental organizations to provide functions for the good and welfare of society. Thus, a stated goal of U.S. governmental technology policy is to make the most efficient use of taxpayer-supported science and technology in achieving national security and the national goals of increased U.S. competitiveness in world markets, improved quality of life for all Americans and continued economic growth.

This goal and the political environment in which it is pursued will often cause the management of technology to acquire different aspects from that which occurs in the private sector for commercial purposes. For example, in the interest of national security and long-range economic competitiveness, the government may undertake costly, high risk research and development that is either not affordable by the private sector or represents excessive financial risk.

The importance of cost and risk consideration suggests the following overall structure for considering the management of technology in the federal government:

- Government Funded Mission and Regulatory Organization
- Revenue Generating Government-owned Enterprises

9.1 Government-funded Mission and Regulatory Organizations

The difference from the private sector is that the objectives of the government sector are established apart from the organization and are characterized by the absence of a profit motive. Similarly, governmental budgets for accomplishing these objectives are determined by the legislature as are the staffing levels.

Within the government sector there is variation in technology development. The defense and space organizations (including National Oceanic Atmospheric Administration ("NOAA")) are high technology organizations that manage and implement the full technology cycle. The development cycle encompasses the four phases: basic research, applied research, development, and operations (Prehoda 1967, p. 17).

Basic Research: Discovery of fundamental new principles about natural phenomena is considered basic research. The objective of basic research is to increase the knowledge base of physical phenomena and processes. This phase involves:

- Synthesis of hypothesis

- Theory
- Observation

This activity is usually done without any concern for practical applications. However, without basic research technological progress would soon be reduced to chance. Invention, in some instances, has preceded scientific explanation. Basic research seems to thrive in an advanced academic environment

Applied Research: Known principles are exploited and useful applications can be foreseen. This is the invention and innovation stage and may include technology demonstration. However, applied research has circumscribed and limited goals.

Development: During development, engineers usually replace the scientist in the process. This is the stage where engineering and economics are used to determine whether a particular technological innovation can be *achieved* within economic constraints and where those proposals found to be technically and economically feasible are implemented. This is also known as the *hardware* stage.

Operations: At the outset of this stage the system resulting from the development phase is accepted by the ultimate user who continues to deploy, operate and maintain the system for its intended purpose throughout its lifetime.

In executing this cycle, within the defense and space component of this sector, the government makes extensive use of industrial contracts to bring technology from conception to implementation, with industry playing the dominant role in development. By contrast, the non-defense and space mission and regulatory governmental organization R&D efforts are more concentrated on the pre-development phases of the technology cycle and a significantly higher percentage of the overall effort may be in governmental laboratories using governmental staffing.

One of the important issues is the relationship between defense and space spending and economic performance. Defense and space governmental R&D programs have supported basic and applied-research studies along with applied and development activities. Direct and indirect linkages appear to exist between defense and space R&D, technology and economic performance (Chakrabarti and Anyonwu 1993). These include:

- direct effects on private sector technology, technical-skills formation, and overall economic performance;

- indirect effects on private sector economy via its effects on technical change; and
- indirect effects on the private sector on technical-skills formation.

With the decline of the Soviet threat, the urgency of domestic problems, and the budget deficit, public pressure has intensified for the defense and space research and development community to produce technologies with commercial applications (U. S. Army 1993). Evidence of the importance of governmental R&D on civilian economic performance, in an indirect way, has been provided through the analysis of time-series data on U.S. defense R&D activities by Charabarti and Anyonwu. The effect is observed particularly through change as measured by the number of patents granted to organizations and individuals. Chakrabarti and Anyonwu argue that they believe that patents are the best available indicator for technology. Data also indicates that government promotes R&D investment by awarding major contracts through a competitive procurement process. By this process, the government reveals the demand for certain technological innovations and encourages the private sector to invest in R&D. Chakrabarti and Anyonwu also observed that the non-R&D aspect of defense spending appeared to have *no statistically significant* effect on either of the major components of civilian economic performance, i.e., technological change and technical skills formation.

Increasing the integration between defense, space and civilian technology may lower governmental cost, promote increased private sector use, increase available industrial capacity, and likely strengthen national security. These outcomes are yet to be verified. Nevertheless, the defense sector has placed increased emphasis on research and development aimed at technologies that may enhance industrial competitiveness over the long and short term.

This change of a policy paradigm has been accelerated by a number of factors including the following statement by the Clinton Administration (Clinton and Gore 1993):

> *"American technology must move in a new direction to build economic strength and spur economic growth. The traditional federal role in technology development has been limited to support of basic science and mission-oriented research is the Defense Department, NASA, and other agencies. We cannot rely on the*

serendipitous application of defense technology to the private sector. We must aim at these new challenges and focus our efforts on new opportunities before us, recognize that governments can play a key role helping private firms develop and profit from Innovation."

This raises certain questions as to the process of trying to refocus defense and space technological activities and how these activities are managed.

9.2 Revenue Generating Government Owned Enterprises

While units such as the U.S. Postal Service, Uranium Enrichment Corporation, municipal utilities and other Government Owned Government Operated ("GOGO") organizations do not have a profit incentive similar to the private sector, they do have measures of economic effectiveness, and competitors. In this sense, the management of technology within these organizations is similar to the private sector.

10. MANAGEMENT OF INTERNATIONAL TECHNOLOGY

The economy of the world is being transformed by a technological revolution. The transformation emanates from the information technology section (Preeg 1994). The transformation is international in nature and therefore much of the adjustment in national economies is transmitted through the international trading system. Comprehensive data is not available that identifies international technology flows. Much of this trade is of R&D intensive exports, in which technology is embedded.

It is therefore not surprising to recognize that technological developments have major impacts. The impacts on international policy in both economic and political-military spheres can be substantial. Technology moves across national boundaries in a number of ways (see Chapter Seven *Technology Transfer*):

International Technology Market: Independent buyers and suppliers of technology utilizing conventional economic channels of transfer such as sale and delivery.

Intra-firm Transfer: Technology transfer takes place through either a joint venture or wholly owned subsidiary, becoming increasingly popular in transfer of technology to developing nations.

Government Agreements and Exchange: Counterparts can be either public or private actors who transfer technology through Co-operate Research and Development Agreements ("CRADA"), and other instrumentalities.

Education, Training and Conferences: Dissemination of information is made public for common consumption by either a general or specialized audience by participation of technologists who present and are present at these sessions.

Pirating or Reverse-engineering: In a number of cases, access to technology is obtained while resort to market is avoided, but at the expense of the proprietary rights of the owner(s) of the technology through knowledgeable non-authorized use.

The management of international technology exists when the production plant is located in one country and some of the technological and managerial inputs to the investment process are imported from suppliers in another country. Similarly, the technological process may all be located in one country but the market is world wide. Many multinational companies have plants in various countries with management and design functions in other countries. Communication by advanced communication systems such as wide area networks ("WAN") provide the means for the enterprise to effectively operate.

According to Kenichi Ohmae (Ohmae 1990, p. 17), most managers in companies that have operated internationally for years are *nearsighted.* Accordingly these managers may manage complex organizations with elements in a number of different countries, or have joint ventures, sources and sell all over the world, but their vision is usually limited to their home-country customers and organizational units that serve them.

Additional talents and background are required in international technological management in comparison to enterprises dealing primarily in one country -- a rapidly decreasing number. These characteristics include:

- knowledge of international trade and the regulations covering these trades;
- knowledge and sensitivity of the socio-economic environments in which they will operate;
- knowledge of international financial instruments, currency and other economic considerations; and

- knowledge of the legal systems in which the enterprise will operate.

There are various elements of the management of international technology where one country supplies the technology and the other produces, according to Quazi (Quazi 1995):

Management and execution of R&D: This is performed by the suppliers of the technology.

Management and execution of pre-investment and feasibility study: This task is performed either by the technology importer or consulting firm(s) hired by them.

Management and execution of design and engineering services: These services are usually performed by the technology suppliers.

Management and execution of capital goods production: Process management and total involvement of the employees are directly applicable to this phase. If the suppliers and the purchaser work as partners, better outcomes are expected as the purchaser has better knowledge of the local conditions including the needs and expectations of the ultimate consumer.

Management and execution of installation and commissioning service: Leadership, effective communications, education and training and process control are the important elements in this stage of the process.

REFERENCES

Betz, F. (1987). *Managing Technology*, Prentice-Hall, Inc., Englewood Cliffs, NJ.

Bright, J. R. (1968). "The Manager and Technological Forecasting." Technology Forecasting for Industry and Government, J. R. Bright, ed., Prentice-Hall, Inc., Englewood Cliffs, NJ, 343-369.

Cardwell, D. (1994). *The Norton History of Technology*, W. W. Norton & Company, New York, NY.

Chakrabarti, A. K., and Anyonwu, C. L. (1993). "Defense R&D, Technology and Economic Performance: A Longitudinal Analysis of the U.S. Experience." *IEEE Engineering Management*, 40(May 1993), 136-145.

Clinton, W. J., and Gore, A., Jr. (1993). "Technology for America's Economic Growth: A New Direction to Build Economic Strength." , Office of the President of the United States, Washington, DC.

Economist, T. (1996). "Semiconductor: When the chips are down." The Economist, 19 - 21.

Edosomwan, J. A. (1989). *Integrating Innovation and Technology Management*, John Wiley & Sons, New York, NY.

Flood, R. L., and Jackson, M. C. (1991). *Creative Problem Solving - Total System Intervention*, John Wiley and Sons, Chichester, UK.

Freeman, R. C. (1982). "The Future of Peripheral Data Storage." *Mini-Micro System*(February 1982), 175-179.

Gillooly, B. (1995). "PowerPC Shorts Out." Information Week, 24.

Gover, J. E. (1993). "Analysis of U.S. Semiconductor Collaboration." *IEEE Transaction on Engineering Management*, 40(May 1993), 104 - 113.

Grindley, P., Mowery, D.C., and Silverman, B. "SEMATECH and Collaborative Research: Lessons in the Design of High-Technology Consortia." *Journal of Policy Analysis and Management*, 13(Fall 1994), 723-758

Hamel, G., and Prahalad, C. K. (1994). *Competing for the Future*, Harvard Business School Press, Boston, MA.

Jabbar, M. A. "Globalisation of Disk-Drive Industry." *IEEE Annual International Engineering Management Conference*, Singapore, 122-127.

Jantsch, E. (1967). *Technological Forecasting in Perspective*, Organisation for Economic Co-operation and Development, Paris, France.

Lerner, E. J. (1994). "Architecting the PowerPC." *IEEE-Engineering Management Review*, 22(Winter 1994), 8-11.

Martino, J. P. (1993). *Technological Forecasting for Decision Making*, McGraw-Hill, Inc., New York, N.Y.

Microsoft, C. (1995). "Encarta'95." , Microsoft Corporation, Redlands, WA.

Ohmae, K. (1990). *The Borderless World: Power and Strategy in the Interlinked Economy*, Harper Business, New York, NY.

Porter, A. L., Roper, A. T., Mason, T. W., Rossini, F. A., and Banks, J. (1991). *Forecasting and Management of Technology*, John Wiley & Sons, New York, NY.

Preeg, E. H. (1994). "Who's Benefiting Whom? - A trade Agenda for High-Technology Industries." *IEEE-Engineering Management Review*, 22 (Summer 1994), 75-83.

Prehoda, R. W. (1967). *Designing The Future - The Role of Technological Forecasting*, Chilton Book Co., Philadelphia, PA.

Quazi, H. A., "Application of TQM Principles in International Technology Transfer Process: an Integrating Framework." *IEEE Annual International Engineering Management*, Singapore, 128-133.

U. S. Army, (1993). "Army Science and Technology Master Plan." *vol. 1 and 2*, U. S. Department of Defense, Washington, DC.

DISCUSSION QUESTIONS

1. Use the printed and electronic media, such as Internet, clip articles dealing with technological advances. Analyze these articles by describing the technology prior to the advance, the advance itself and the technological capability after the advance.

2. Prepare a listing of the pros and cons of the following statements:

   a. Technological change is not necessarily progress.

   b. Technology strengthens the power of an elite *technic* minority within society.

    c.   Technology developments need not be based upon basic scientific developments.

    d.   Technological progress leads to economic progress.

    e.   Technology is driven solely by societal needs.

    f.   Technology is driven solely by competitive needs.

3. Review the history of television distribution technologies.

    a.   What are the major distribution systems?

    b.   What are the advantages and disadvantages of each?

    c.   What scientific technological events are driving the choices and estimate which technology will eventually predominate and why?

4. Trace the history of the development of Internet from a government sponsored system to its current state of commercialization.

    a.   Determine the technologies which caused the growth of Internet, both hardware and software.

    b.   What are the characteristics of the companies who are the leaders in the Internet system?

5. Comment on a technology for which there is a steady demand and that receives broadbased support from a wide variety of sources, i.e., private industry, governmental and multi-national groups, each seeking competitive advantages by an advance in the level of functional capability. Present this in terms of organizations and their management of the technology development.

6. The Novell Corporation has been a leader in network development. Comment on the past, current and potential future of this company in terms of its management of technology.

7. Chose an organization which you are closely familiar with and discuss to the best of your knowledge how it manages technology. Show how it interacts in its sector, i.e. private, governmental or international.

CHAPTER 2

Technological Strategy

1. INTRODUCTION

What is strategy and, in particular, what is technological strategy? Once these questions are answered, we are still faced with the problem of the formulation of technological strategy within an enterprise. This chapter will try to present the answer to these questions and provide a methodology for formulating a rational technological strategy within an enterprise.

Throughout history, various groups and individuals have developed strategies to reach their goals and objectives. However, in many instances the roadmaps that they have laid out for themselves and their enterprises have led to unexpected results. Having goals and objectives is an important element on the path to their achievement, but these must be coupled with the roadmap or strategy for moving toward these goals and objectives. While a methodology for developing roadmaps or strategies can be presented, it must be clearly understood that the environment, through stochastic resonance, (see Chapter One - *Technological Advancement and Competitive Advantage*) can move individuals and enterprises toward unexpected results. Stochastic resonance, coupled with cascaded events, are the culprits in many failed technological strategies.

An example of this combination of environment and poor technological strategic planning is the failure of Apple Computer, Inc. to become the dominant computer enterprise, and achieve a technological lockout of other alternatives. While Apple's technology was the undisputed leader in *user friendly* systems, its strategy of maintaining a closed system eventually caused secondary technologies to penetrate and eventually dominate the market. The failure of Apple's technological strategy can be attributed to the environment that initially launched the enterprise that became Apple Computer, Inc., i.e., the mid-1970s populist view of technology of Steven

Jobs and his associates. In the late 1980s Apple Computer had introduced the Macintosh® computer, its object centered operating system. Apple overcame many of its initial problems, and had the potential of becoming the dominant technology. When Apple introduced its Macintosh® computers, the IBM compatible personal computers ("PCs") were user-hostile and relied on a difficult and confusing disk operating system ("DOS"). By the late 1980s John Sculley, a marketing executive who had become Apple's Chief Executive, had started to impose his strategic vision on Apple Computer. Mr. Sculley kept Steve Jobs' original populist vision and superimposed Mr. Scully's marketing based vision on Apple Computer. This strategy relied on keeping prices high and refusing to license the Apple user friendly operating systems to other manufacturers. This opening of use by other manufacturers of the user friendly operating system would have led to the development of low-price alternatives to the Macintosh® and possibly likely greater market penetration and market dominance. This coupling of strategies by Mr. Sculley to maintain Apple's monopolistic position eventually led to its loss of market position by others taking advantage of technological advance that was not fully exploited.

Microsoft®, recognizing the appeal of the object centered operating system of Apple, launched its Windows® program which, by then, operated on lower-priced IBM clones. This led to *technological lockout* of Apple Computer in the PC market place. A closed operating environment led to an increasing entropy situation, i.e., reduced profitability and possibly eventual a form of *thermodynamic* death for Apple Computer; this is an extension of Ilya Priogogine's work on entropy and documented by Çambel (Çambel 1993, p. 136).

1.1   The Meaning of Strategy

A *strategy* is the pattern or plan that integrates an organization's major goals, policies, and action sequences into a chosen whole (Mintzberg and Quinn 1991, p. 5). A strategy serves to allocate an enterprise's resources into a unique and viable posture which is based in the organization's:

- core competencies and shortcomings,
- unanticipated changes in the environment, and

- contingent moves by competitors or *agents*[1].

One of the basic structures used to view strategy was developed by Mintzberg (Mintzberg and Quinn 1991). A model for an overall corporate strategy is shown in Figure 2.1. Figure 2.1 is in comparison of how a strategy impacts the enterprise given by Pearce and Robinson and shown in Figure 2.2.

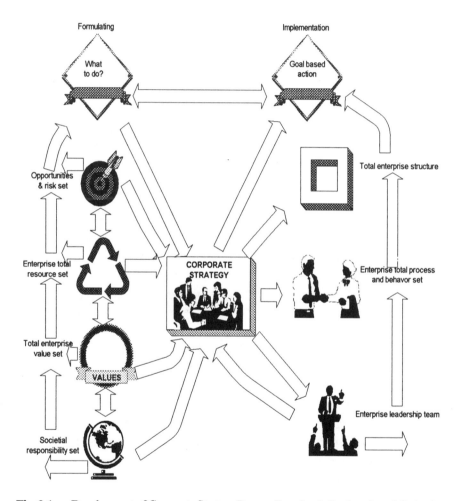

Fig. 2.1   Development of Corporate Strategy Process Based on Mintzberg's and Quinn's (Source: Mintzberg and Quinn 1991)

---

[1] What is meant by an agent is either a process or series of processes launched adversely or unintentionally by nature or other components of the environment.

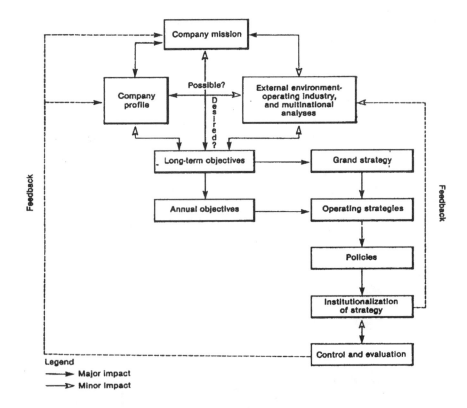

Fig. 2.2 Pearce's and Robinson's Strategic Management Model (Source: Pearce and Robinson 1991, p. 12. Reprinted by permission of Richard D. Irwin, Inc.)

Mintzberg and Quinn (Mintzberg and Quinn 1991, p. 11) suggest that effective strategies should, at a minimum, encompass certain critical factors and structured elements such as:

- *Clear, decisive objectives* -- Effects directed toward clearly understood, decisive, and attainable objectives.
- *Maintaining the initiative* -- The strategy preserves the freedom of action and enhances commitment.
- *Concentration* - Concentrates power at the place and time likely to be decisive.
- *Flexibility* - Purposefully built in resource buffers and dimensions for flexibility and maneuver.

Strategies are not independent of the *nested* set of functions. One way to view strategy development is to use a *Top Down* approach. This methodology allows an enterprise to logically build upon requirements to satisfy the enterprise mission. The *Top Down* approach allows managers of technology to provide a roadmap for the *multifunctional teams* (see Chapter Four) during the development process. The *Top Down* methodology can be viewed as a pyramidal structure (see Figure 2.3) and consists of:

- Vision
- Objectives
- Goals
- Policies
- Programs
- Strategic Decisions.

Fig. 2.3 Strategic Top Down Methodology

To place the methodology in perspective we should use the same definitions for the functions since the meaning can greatly influence the final outcome, i.e., a consistent and *vectored* set of enterprise activities.

1.1.1 Vision

Vision is the term used to imply the general statement of the direction and potential end point for an enterprise, usually stated by the chief decision maker ("CDM"). This statement of desired direction, or *organizational vector*, may or may not be based on any informed foresight. Napoleon's vision of a United Europe, while correct, was not able to be achieved, due to the existing social or economic or political environment, while Bill Gates's vision for Microsoft®, as monopolistic operating system, was capable of achievement within the existing market environment augmented by *stochastic resonance.*

1.1.2 Objectives and Goals

The word *objective* is defined by <u>The Concise Oxford Dictionary</u> as:

> *"objective n*
> *1 something sought or aimed at; an objective point.*
> *goal n.*
> *2 the object of a person's ambition or effort; a destination; an aim (fame is his goal; London was our goal).*
> *3 a point marking the end of a race."*

As these definitions illustrate, confusion can arise between the terms *objectives* and *goals*. Organizational writers have taken different approaches to contrast the two terms.

Essentially, *objectives* and *goals* both specify end points sought, but they differ with respect to the time period and the degree of specificity. *Strategies* refer to means of attaining these broad results sought. In either instance one or the other represents a nested set, i.e., in terms previously defined, goals are nested within objectives.

This text will use the definition that the *objectives* represent a broader term purpose with longer term implications, i.e. the future position, while *goals* represent a more focused result to be achieved in a shorter time frame, i.e., a near term enterprise position. Some organizations such as United States National Aeronautics and Space Administration ("NASA") reverses these concepts, i.e., *goals* are broader than *objectives.*

In organizational terms, *objectives* are what the firm or organizational unit attempts to accomplish in the long term. *Objectives* for the firm and its units tend to be broad, general statements. In an operations research ("OR") sense, *objectives* mean the results that the decision-maker wants or should want in regard to a particular system or problem. There is a definite relationship between the hierarchy of factors that influence decisions and decision-makers. These factors include the mission, objectives, strategy, goals, programs or projects, and resource allocation of the enterprise and can be summarized as:

- *Mission* -- the *business or activity area* that the enterprise is engaged in

- *Objectives* -- desired future positions or roles for the organization, i.e., long term resulting position
- *Strategy* -- the *general* direction in which the objectives are to be pursued
- *Goals* -- specific targets to be sought at specific points in time, near term
- *Programs or projects* -- activities consuming sets of resources through which strategies are implemented and objectives pursued.
- *Resource allocation* -- allocation of funds, personnel, etc., to various units, objectives, strategies, programs and projects.

The decision involved in setting necessary *objectives* is a fundamental responsibility of top-level management. *Objectives* are hierarchical in nature in the sense that the general overall *objectives* are formulated by top management after economic, social, legal, technical and political forces (i.e., environments) affecting the organization have been appraised. These overall objectives provide guidelines for the organization and become the standard against which progress can be measured.

1.1.3 Programs

Programs should derive from the *objectives* of the organization. The basic difficulty involved in defining appropriate programs has to do with the nebulous objectives of most large organizations. Also, in government and educational organizations, the objectives are not usually operationally defined, i.e., defined in a way in which attainment can be readily measured.

Because objectives are often vaguely stated, the relationships between the organization's activities and its objectives are seldom precisely understood. However, the organization knows, or believes that it knows, those areas in which achievement will lead to the attainment of its objectives. Such a non-rigorous relationship between outputs and objectives is often an adequate basis for the definition of a meaningful program structure for an organization.

Traditional management usually had *unilateral objectives*, i.e., it tends to emphasize the satisfaction of a single objective such as the stockholders' claim and profitability. The objectives of today's enterprise extend to a range of *clientele or customers or stakeholders*, each having its parochial goals. The challenge for today's management is to maintain overall organizational effectiveness in the presence of multiplicity of sometimes conflicting objectives.

The development of objectives at one organizational level that can be broken down into sub-objectives for lower organizational levels provides a concrete beginning point for the development of organizational plans. Consistency is the *sine qua non* of successful organizational plans.

The process of *successful* organizational management requires:

- A common set of shared objectives and beliefs among organizational leadership.
- Communication of these objectives and beliefs to the performers, i.e., workers.

Without broad, stable and ongoing objectives, managers and performers (i.e., workers) have no way of knowing what to do, or why they should do anything. Objectives can be pursued in different ways and approached by different paths. To co-ordinate organizational activities in pursuit of goals, organizations develop strategies to indicate the path management wants to follow to achieve its objectives. The strategies that are used indicate the paths and *velocity of technological movement*, i.e., *technological vector*, to be followed. These strategies include policies, general guides for acting, and program of activities for realizing goals. The process of selecting objectives and developing programs involves strategic planning.

Organizational objectives appear in different forms, and often conflict with each other. These conflicts need to be resolved in some way so the organization can function as a co-ordinated whole. When conflicts between goals are very extensive and reduce organizational effectiveness, then it may be necessary to separate the organization into separate components so that co-ordinated objectives can be established for the various separate entities.

An organization's objectives should be reviewed from time to time to determine if they have some validity, as when they were first formulated. Care should be taken in changing them for the disruption of the system could lead to instability.

It is important in management control systems to emphasize the distinction between objectives and goals. To confuse the two ends could misdirect the flow of control information to different levels of management and might motivate senior management to save the tree and lose the forest.

1.1.4   Policies

Policies are the rules or guidelines that express the limits within which actions should occur.   In other words, policies should be a consistent set of constraints in which a strategy is implemented.   *Strategic Decisions* are those decisions that determine the overall direction of the enterprise and its ultimate viability in light of the stochastic nature of the environment.

1.2   Types of Strategy

According to Mintzberg there are a number of different types of strategies that can be pursued by an enterprise (Mintzberg and Quinn 1991, p. 15):

Planned Strategy: A highly deliberate set of precise intentions that are formulated and articulated by the Chief Decision Maker ("CDM") or leadership group within a formal control structure to insure surprise-free implementation under a high probability environment.

Entrepreneurial Strategy: Deliberate but potentially emerging set of personal, unarticulated visions of a single CDM that are adaptable to new opportunities.   The enterprise is under the control of the CDM and is in a niche environment.

Ideological Strategy: Rather deliberate set of intentions shared by all elements of the enterprise, controlled through strong shared norms and proactive vis-à-vis its environment.

Umbrella Strategy: CDM or leadership group in partial control of the enterprise actions defines strategic targets or boundaries within which others must function.   As a result, strategies are partly deliberate and partly emergent.   This allows others, with leadership approval, the flexibility to maneuver and form patterns within boundaries.

Process Strategy: This partly deliberate (process) and partly emergent (content) strategy is achieved by the CDM/leadership group that controls the process aspects of strategy, leaving the actual content of the strategy to others.

Disconnected Strategy: Members or subunits who are loosely coupled to the rest of the enterprise produce patterns in the streams of their own actions in the absence of, or in direct contradiction to, the central or common intentions of the enterprise at large.

Consensus Strategy: Emergent set of strategies that are produced through mutual adjustment by various enterprise members who converge on patterns that pervade the organization in the absence of central or common intentions.

Imposed Strategy: Enterprise emergent strategies, which may be internalized and made deliberate, derived from external environmental actions, either through direct imposition or through implicitly pre-empting or bounding enterprise choices.

The strategies in this classification, while relatively global, are by no means the only forms of strategy. It is possible to postulate another set of strategy definitions which will contain various elements of Mintzberg's classification discussed above. In many instances, organizations develop strategies because of numerous internal and external factors. Mintzberg's classification is for *pure* strategies, i.e., a strategy with only one set of underlying factors.

1.3   Dimensions of Technological Strategy

Formal strategies have been used in many contexts over the centuries. It is possible to draw certain conclusions from these prior strategies and apply them to technological strategy. Any formal strategy must contain several essential elements. An overriding and obvious element is an objective of the strategy which is desired to be achieved. The structure for implementation is the policies which serve as the guide and constraints during implementation.

A technological strategy, paraphrasing Betz, Martino and Mintzberg among others, can be stated as:

> *a formal set of enterprise technological*
> *intentions that allocates available resources*
> *and sets priorities based on clearly stated*
> *technological and enterprise objectives and a*
> *perceived environment in which the process*
> *is to be embedded.*

The dimension of a formal technological strategy expands these essential elements and adds several other important considerations. These additions include: key technological concepts and thrusts (*technological vector*); structure for unpredictability and unknowability of the environment (*stochastic elements*); and being part of a total set of enterprise consistent strategies (*total enterprise strategic system.*)

Without a formal technological objective, which is nested within and supportive of the enterprise's system of objectives and mission, no formalized technological strategy will succeed. Effective technological strategies develop because of a few key technological concepts or thrusts (*enterprise technology vector*) which form the nuclei giving them cohesion, balance and focus. Sufficient enterprise resources must be allocated for success of the strategy. This means that all elements of the enterprise must be co-ordinated and actions controlled to support the total technological strategy. A possible, but not yet proven example, is Microsoft's® development of Windows'95™. Microsoft® has structured their strategy so that all elements support a thirty-two bit architecture with both backward and forward compatibility. Their technological vector leads to a potential enterprise dominance through incremental improvement. However, the rapid rise of Internet and distributive software such as Sun Corporation's Java™ have caused Microsoft® to re-evaluate their technological strategy. Organizations that do not re-evaluate strategies after the introduction of competing technologies can suffer the fate of companies such as Wang Laboratories which did not adjust strategies during the rapid growth of the PC.

A technological strategy deals with both the unpredictable and the stochastic. The potential stochastic resonance, both positive and negative, of the environment must be dealt with by a technological strategy. This includes internal enterprise developments, and this is especially so in a large organization. Large-scale (*non-linear*) systems can respond quite counter-intuitively to apparent rational actions. Small changes can drive these systems unstable (*chaotic*). Consequently, the technological strategy is to build a strong and flexible strategy that can achieve its objectives despite the unforeseeable warp the external environment may impose on the strategy.

Technological enterprises should have hierarchically related and mutually supporting strategies. Each of the strategies, i.e., technological, market, and enterprise, while more or less complete in themselves, must be a cohesive element of higher level intentions of the organization and form a co-ordinated system.

1.4   Technological Strategy Formation Process

Technological strategy formulation is a long-range continuous process. Developments in the technological arena are both rapid and far-reaching in extending or contracting opportunities for an enterprise. This has considerable impact when the principal function of the enterprise is technology. Technological change can alter the fundamental nature of an enterprise and its activities. These developments and enterprises must be captured during the technological strategy formation process.

As previously stated, a technological strategy is the general vector in which the technological objectives of an enterprise are to be pursued. The term *vector* is used to imply direction and magnitude. The formation of a technological strategy must take a dynamic view of the future environment. The strategy must provide the direction, and emphasize the magnitude of the technological process which will be needed to achieve these objectives. Technological strategy cannot be developed in isolation; it should be an integrated part of overall strategic enterprise planning activities. This linkage, according to Betz (Betz 1993) relates to future arenas of enterprise activity, i.e. markets or competitive enterprise activity theaters. Positioning of enterprise in these arenas and the necessary research, production and marketing to attain the desired position is another of the linkages. In military terminology, choose the battlefield, position resources to maximize capabilities and strike at the enemy's weakest position.

There are three organizational considerations that must be incorporated into a technological strategy. These considerations start with desires of the enterprise, which must be matched by the technological capabilities of the enterprise, and in turn match the needs of the market place. The difficulty with technology is that market needs change rapidly. It is possible that the technology might have been the right product when the technology was conceived, but market forces may find it totally irrelevant at the time the technology is finally available for implementation. This is very often seen in the information industry where market needs change rapidly.

An enterprise's desires have usually been formulated by the CDM or the leadership team. The desires, in some instances, are not formalized but exist within the organization. The statements and actions of the CDM or leadership team usually contain elements that can be extracted to develop what the true desires of the organization are in relationship to the overall interests and technology strategic vector. The organizational desires, set by the CDM or the leadership team, formally or informally, are a form of a constraint on the eventual implementation of any strategy.

The capabilities of an enterprise are the total set of resources that are available to implement any strategy, and form a set of constraints for implementation of any strategy. In the implementation of a technological strategy the technical capabilities of an organization may need further augmentation to fully implement the developed strategy.

The market needs represent the environment that the technological strategy must satisfy to meet the enterprise's technology objectives. In some instances, the market needs may not easily be fully known. New technologies sometimes foster new needs. An example is the PC that did not exist in a personal environment prior to 1970. These unrecognized needs are the area which many entrepreneurs appear to focus upon, and in many instances they anticipate market needs many years before those needs become visualized within an economic structure.

Technological strategy cannot be developed without consideration of the overall enterprise strategy. Therefore an important element is to develop an integrated technological strategy that assists the enterprise's overall competitive position. This requires managers of technology make sure they fully understand the vision and objectives of the enterprise's leadership team.

Technological strategy formation must also consider all phases of the value chain of an enterprise. Porter's general approach applies to formulating technological strategy consisting of a number of the generic elements modified for use within a technological enterprise (Porter 1985).

It is, first, important to identify all technology categories in the value chain for the enterprise. These are not only the initial technology categories which are inherent within the development, but technology categories which are ancillary, such as manufacturing and service categories which will become prominent once the development enters its intended market

Within these categories, it will be important to identify relevant technologies and trends. Technology vectors will determine the ability of the enterprise to successfully implement the intended strategy. In a number of instances, developments have been undertaken when one or more of the important technologies was not sufficiently mature to meet market requirements. Apple's Newton® Personal Digital Assistant ("PDA") faced this due to the lack of maturity of handwriting recognition software development.

Therefore, it is important to determine which technologies and potential changes are most significant in the strategy formulation. It is possible to dissect a potential development in a form similar to a branching network. The branches represent the various technologies associated with the development and alternatives. Using network analysis it is possible to determine the impacts on a technological strategy being proposed.

Any technological strategy will require the utilization of various resources. Thus, a manager of technology must assess resource requirements, costs, and associated risks. The manager must also determine alternatives for improvements in relevant technologies.

Once the technological strategy has been developed, it should be applied consistently throughout the enterprise. This requires that the leadership team of the enterprise fully accept and back the implementation of the technological strategic plan throughout all elements of the enterprise. Thus, communication is a critical element in the process. This communication should start with the beginning of the design of the strategy and be an ongoing process.

Porter only provides a general approach to the development of formulating a strategy which can be used in a technological enterprise. There are a number of techniques that have been developed to produce a technological strategy. These techniques include:

- *Scenario Analysis* : This process can provide an effective format for determining the impact of technological change in an enterprise's activities (Kahn and Wiener 1967).
- *Technological Mapping*: Another technique is Pyke's (Pyke 1973) hierarchical mapping approach.

1.4.1 Scenarios

A scenario is a written description of situations popularized by Herman Kahn in the 1960s. According to Kahn and Wiener (Kahn and Wiener 1967):

> *"Scenarios are hypothetical sequences of events constructed for the purpose of focusing attention on causal processes and decision-points."*

Scenarios answer the question how a hypothetical situation would develop, step by step. Thus the developer of the scenario must assume a number of branching situations. This requires making assumptions on what alternatives could exist, for each actor, at each step, for presenting, directing, or facilitating the process. There are a number of steps in writing a scenario (Martino 1993, p. 215-220). These steps begin by first developing a framework for the scenario.

Scenario Framework Development

The framework for any scenario is a set of variables which set the stage and, like, any theatrical production, should be as realistically defined as possible.

Environment: The environmental variables describe what occurs in the various external components which can impact the scenario and could include societal, market, competitive, political and technological environments in which the technology is to be embedded. In many instances, it is stochastic and uncontrollable environments which, in large measure, will determine the eventual success of any strategy. In describing the environmental elements, the scenario builder should not be overly optimistic, since it is unlikely that such environmental assumptions will result in a realistic scenario.

Trends: It is important to know what are the important trends than can influence the outcome, the direction and the continuity of a technological strategy. This is not necessarily a trivial exercise since it relates to having realistically analyzed the previously discussed environment. However, a number of analytical tools are available (see Chapter Three - Technological Forecasting).

Societal choices: How societal choices impact the technological forecast must also be considered. The societal choices are vectorial and have magnitude and direction. However, some critical events, such as the Three Mile Island nuclear plant hazardous nuclear release, can sensitize society in a manner which substantially changes these vectors.

Critical decisions: How the societal choices influence critical decisions and when these decisions should be made is also an important element. The "right" decision at the "wrong" time will, in many instances, have the opposite impact to that desired. An example is governments having failed to release critical data for fear of societal reaction. When this data has been eventually disclosed, the societal reaction has been much more severe then if the data had been disclosed when an event happened. The Chernobyl reactor accident, in the former Soviet Union (Ukraine), was an example of a technological disaster which, due to the withholding of critical information, helped accelerate the collapse of an empire.

Chief Decision Maker ("CDM"): Any scenario must clearly define the CDM who is responsible for making the critical decisions. The characteristics and heuristics of the CDM are another important consideration in the development of a realistic scenario. Literally, a psychological profile of the CDM would be useful in developing the technological scenario. This would allow the scenario builder the ability to make more realistic judgments of the reaction of the CDM to various environmental changes which can occur.

Technological Forecast Development

Once the stage has been set, the next step is to forecast the technology or technologies to be considered (see Chapter Three - *Technological Forecasting*). In general, there a number of steps used in forecasting technologies.

Deployment: The timing of the deployment of the technology is one of the first items be decided. The means making assumption as to the time necessary to develop, and then deploy the technological development within the intended environment. If the assumption is overly optimistic, it is possible that competing technological developments will not be available when the selected technology is to be deployed. In fact such an assumption could make developers overly confident and more likely to miss important developmental steps which add additional advantages. Similarly, being overly conservative on timing can lead developers to prolong development beyond the point where the marginal benefits do not justify the loss of competitive advantage.

Scale of deployment: A technology can be deployed from a trial test mode to full implementation. Any developmental program goes through various phases in the technological life cycle (see Chapter Five - *Technological Life Cycles and Decision Making*). The earliest deployment will likely be trial test, "*Beta Testing*". Additional trials and revisions can also be part of the deployment cycle, until the final development is completely deployed.

Time from adoption: Most technologies require time to penetrate the target market. Any technological forecast must consider this timing factor. Past history for similar developments can serve as a staring point to determine market penetration and penetration rates.

Impacts of technology on trends, events in framework and commensurate impacts of the framework on technology: The system that is being considered is highly complex and interactive. The interaction between the technology and framework, including potential cascading events, is an important aspect of the technology forecasting process.

Critical technological decisions: Critical decisions are those decisions which can have a significant impact on the technology. There are a number of critical decisions which should be specified during the implementation of technology. These decisions form the basis for future actions. The use of decision tree methodologies (see Chapter Five - *Technological Life Cycles and Decision Making*) is a useful approach to developing a critical technological decision framework.

Timing of critical decisions: No decision is made in an abstract time frame. It is important that the timing of critical technology decisions in the scenario be carefully considered. This is an important factor since these decisions will interact with various events occurring within the environments under consideration.

Decision makers for critical decisions: All enterprises have decision makers at various organizational levels: a developmental team leader, a Chief Technology Officer ("CTO"), a Chief Executive Officer ("CEO"). Each level of a technological enterprise has different responsibilities and decision making authorities.

Scenario Preparation

After the framework technological forecast have been developed, the next step in the process is the preparation of the scenario or a group of scenarios. The first step in the process of preparing a scenario is the designation of a sequence of events and decisions. The manager preparing the scenario identifies the key decision trigger events. It is important that the sequence of events and decisions be a consistent set if the scenario is to have any validity.

The next step in the process is to prepare the written scenario or group of scenarios. The general approach is to take the framework previously developed and fill in the narrative describing the events. According to Martino (Martino 1993, p. 218-219), there are several approaches to writing scenarios. The process is similar to the approach taken by authors of fiction in preparing novels and theatrical productions.

Looking Backwards: In this approach, which was popularized by Edward Bellamy[2] in the nineteenth century, the scenario is prepared from the perspective of a specific time in the future, after events have taken place ~ *"how we got where we are."* The advantage of this Bellamy approach is that it gives a global view, i.e., a holistic approach, with a macro perspective. The

---

[2] **Bellamy, Edward**, 1850–98, an American author; b. Chicopee Falls, Mass. He became famous with his novel *Looking Backward, 2000–1887* (1888). This was a utopian romantic novel of the future under state socialism.

62

scenario is *broad brush* and thus is not encumbered by micro-details which can obscure important decision points. The disadvantage of this scenario view is that it *fails to get readers involved*. Like any historical approach, it is difficult for the reader to be totally involved in the scenario. The writer of the scenario must insert drama by describing fictitious events, but this causes possible problems. Catastrophic atmospherics can obscure the underlying rationale.

Viewpoint Character: The scenario is written from the viewpoint of individual who is seeing, undergoing or taking part in the events as the scenario unfolds. This device is one of the major approaches used by authors of fiction. The reader becomes involved as a virtual participant and forms a subconscious intuitive relationship with the unfolding events and characters. A personal impact of portrayed events is developed through this process. In this type of scenario, it is difficult to present an overall perspective in the sequence of events and to use this method if it covers a long period of time.

God's-eye view: This is similar to the *viewpoint character* approach; however, the perspective is much more global. It presents the scenario as a series of viewpoints, each from the perspective of a viewer who is seeing history unfolding and reacting to it. The disadvantage in using this method is that it may be a dramatic failure, since there is no one person seeing or telling the story. It also fails to bring the reader in to the story.

Diary: In this case, the scenario appears as a series of diary or journal entries, or extracts from personal letters, written shortly after the events described. This personal approach presents the reader of the scenario a means for *"bonding"* with the scenario writer. A diary describes events as perceived, and includes information not available at the time of the events, such as other associated events. The matter-of-fact nature of diary entries lends verisimilitude to what otherwise would be considered an incredible tale. However, diary entries tend to filter out the drama of the situation and present a sanitized version of events.

A scenario plot outline can only be turned into a full scenario by selecting one of these approaches and actually writing out the events in the chosen form.

1.4.2   Technological Mapping

Technological mapping (Pyke 1973) is a technique in which technological change is systematically analyzed in three hierarchically organized categories:

Level 1 -- Environmental Scanning and Analysis
Level 2 -- Technological Capability Analysis
Level 3 -- Product Line Analysis

These three levels form a nested set of activities in the formation of a technological strategy. The information derived from a particular level serves to assist in providing the next level of analysis.

Level 1 – Environmental Scanning and Analysis

As Chapter One illustrates, there are many precursor events that would have made it possible to forecast the eventual development of a particular technology. The systematic environmental monitoring requires that precursor events be identified. These events are then used to provide the forecast of a potential technological development. The processes used during this level include collection, screening, and analysis.

Collection: The collection process brings environmental information into the system for processing. Collection requires determination of the sources of information that should be monitored.

Screening: The only items used for analysis should be those that are relevant to the formulation of the strategy. Information overload can be a detriment to the process. However, not all environmental information will assist in the formulation of a technological strategy.

Analysis: The information is analyzed to evaluate its significance, i.e., what does this mean to the enterprise. This analysis relates to:

- Impact on mission
- Impact on current operations
- Impact on customers
- Impact on suppliers

Level 2 – Technological Capability Analysis

According to Betz (Betz 1993, p. 41), in any strategic attitude, three concepts are important:

- Perception
- Competence
- Commitment

The enterprise must assess its core technological capabilities. All enterprises have perceptions about their current and future technological capabilities; some of those perceptions are positive and some are negative.

These perceptions have to be tested as to their realistic nature, i.e., *pipe dreams* or *impossible realizations*. The perceptions will impact the ability of the enterprise to develop a realistic and implementable technological strategy.

Technological competency must also be addressed, i.e., the enterprise's ability to develop technologies to meet both the corporate vision and objectives. This includes identifying additional competencies the enterprise must acquire if it is to pursue a particular set of strategic technological objectives. It also includes determining what additional skills, secondary technologies, technological alliances and intellectual property rights the enterprise should develop or acquire.

The final portion of technological capability analysis is to determine the enterprise's commitment to proceed along a particular *technological strategic vector*. Each technological strategy will have a different set of organizational commitments in terms of resources, risk and determination. This can be achieved by reviewing the past technological developments of the enterprise and what was required to achieve the first result. It is possible to use a ranking system and develop a measure of commitment of an enterprise to proceed along various technological strategic paths.

Level 3 – Core or Product Line

Products, services, systems and other enterprise core functional lines are *"born, they live, and they die,"* i.e., they have definite life cycles (see Chapter Five - *Technological Life Cycles*). It is important to determine the life cycle status, i.e., position, of the current enterprise products or services that will be involved in the technological strategy. It is important to determine the ability to extend the product or service lifetime by a particular technological strategy. A determination must be made of the impact of technological strategies on existing products and services, and the impact on potential technological strategies.

1.5   Technological *SWOT* Analysis

The concept of a Strengths Weakness and environmental Opportunities and Threats ("SWOT") analysis derives from general enterprise strategic planning of Pearce and Robinson (Pearce and Robinson 1991, p. 181-185). A technological SWOT analysis is a systematic identification of the factors which must be considered for technological developments. Figure 2.4 shows a technological SWOT analysis diagram based on the methodology of Pearce and Robinson (Pearce and Robinson 1991, p. 184). This analysis is the precursor to the formulation of a successful technological strategy. The environmental analysis, as previously discussed, provides the information needed to identify opportunities and threats in an enterprise's environment.

Strengths are the technological or other resources, skills or advantages of the enterprise and needs to be served by or expected to be served by the enterprise. These can vary from individuals with very specialized knowledge to existing market franchises, and encompass the total set of enterprise technologically related strengths. Many enterprises underestimate certain human resource strengths, which provide the basis for core competencies of the enterprise.

Weaknesses are limitations or deficiencies in technological or other resources, skills, and capabilities that seriously impede an enterprise's effective performance. All enterprises have weaknesses when viewed in a particular context. However, in some instances these weaknesses are overlooked in optimistic atmospheres surrounding key members of the leadership team who may be vying for internal competitive position.

Opportunities are favorable technological, market or other situations in an enterprise's environment. Opportunities can be found in most enterprise environments. An example was the rapid growth of direct mail sales in the computer hardware markets in the 1990s which have been successfully exploited by various enterprises.

Threats are major unfavorable technological, market or other situations in an enterprise's environment. Threats are key impediments to achieving strategic technological objectives. Many enterprises underestimate the threats posed by new untried technologies. The rapid rise of Internet in the mid 1990s caught many firms unprepared including such a market leader as Microsoft® Corporation.

A technological SWOT analysis is used to aid a technological strategy analysis. It forms a logical framework, guiding systematic discussion of an enterprise's situation and basic strategic technological alternatives. It also provides a dynamic and useful framework for strategic technological analysis.

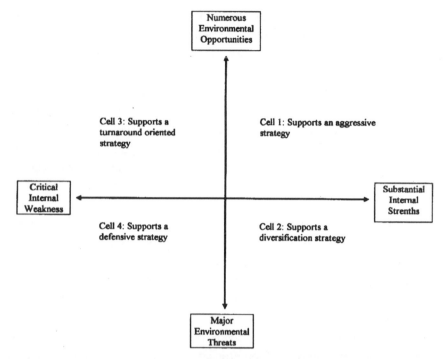

Fig. 2.4 Technological SWOT Analysis Diagram based on Pearce and Robinson approach (Source: Pearce and Robinson 1991, p. 184. Reprinted by permission of Richard D. Irwin, Inc.)

2 TECHNOLOGICAL STRATEGIC PLAN

The steps previously discussed set the stage for the development of a formalized technological strategic plan. The output of any effective planning process is a series of documents called *plans*. *Technological strategic planning* is the ongoing process used by an enterprise to technologically position itself in the complex ever-changing world. This process formalizes the technological strategy that has been formulated into a written plan of action. The process is also an attempt to match internal technological capabilities with external opportunities in the most advantageous way, consistent with long-range organizational goals.

Effective technological strategic planning requires serious and intense organizational involvement to assure reality, performance and commitment (Thamhain 1992, p. 79). Technological strategic planning can also be used to strategize new products or service ventures within the existing set or enterprise portfolio to respond to either a societal or a competitive need.

Figure 2.5 shows the system of strategic plans for an enterprise that are of interest to technology managers.

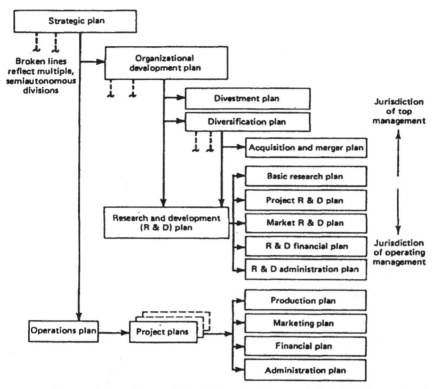

Fig. 2.5 System of Plans (Source: SRI 1963. Reprinted by permission of SRI International, Menlo Park, CA.)

The technological strategic plan should contain a number of topics which clearly define the plan of action of the enterprise in relationship to the technology under consideration. The technological strategic plan at a minimum should contain the following sections:

- Executive summary
- Objectives and goals
- Enterprise
- Technology
- Mission statement
- Environmental analysis
- Existing environment

- Future potential environment
- Technology forecast (see Chapter Three - *Technological Forecasting*)
- Formulation of technological strategy
- Action plan for implementation of technological strategy and associated program elements
- Schedule of actions and milestones
- Resource requirements
- Capital
- Personnel
- Facilities
- Operational budgets
- Organization and delegation
- Applicable policies and procedures
- Appendix of detail supporting facts and assumptions.

The document which presents the technological strategic plan must reflect current and future technological realities as well as organizational desires. The technological strategic plan must also consider and show the interdependencies among various operations such as:

- Research and development
- Engineering
- Production
- Marketing
- Field services
- Finances
- Legal services

The technological strategic plan should also provide a time-phased view of the existing portfolio of products and services. The plan should also show the management effort from its inception to maturity. Important elements, that in many instances are not presented, are the phase-out of the technology and assigned management team. The plan should also contain the:

- Key personnel, including the enterprise's leadership team
- Interfaces both internal and external
- Documentation
- Reviews
- Sign-offs

The technological strategic plan also contains a technology life cycle analysis and other analytical tools (Thamhain 1992, p. 80). These analytical tools assist the strategic planner to help reduce uncertainties to the extent possible in any strategic plan.

Growth-Share Matrix (Henderson 1971) can be used by the technological planner to map the technology life cycle into four quadrants with various degrees of growth potential, market share, and cash flow. Pearce and Robinson (Pearce and Robinson 1991 , p. 262-272) consider this a first step in the strategic plan development.

Profit Impact on Market Strategies ("PIMS") Model (Shoeffler et al. 1974) consists of an evaluation of the data pool of company experiences to determine specific correlation between business strategy measures and profit performance. It has been shown that growth in market share is correlated with profitability. Similarly, increase in markets served, variety of products offered and technologies used also lead to competitive advantages for the enterprise (Pearce and Robinson 1991, p. 59).

Price-Performance Curves (TBCG 1972), a mapping technique, presents the relative value of a technology in relationship to other competitors or technologies.

Experience Curve (TBCG 1972) is a concept for tracking the cost or price of goods and services over time. Plotted on a log-log scale, costs trends become very obvious and are useful in long-range strategy formulation. This is a broader approach than learning curves which present efficiency as a function of time.

Porter Curve shows the generic correlation between market share and new product profitability. The studies of Porter (Porter 1980, p. 43) show that enterprises with very high or very low market share are most likely to be profitable, while enterprises in the middle of the market are more likely to experience lower profitability, and are unlikely to maximize the value their development.

3 APPLICATION OF TECHNOLOGICAL STRATEGIC PLANNING

The following are examples of the application of technological strategic planning by:

- Nations
- Corporations

- Entrepreneurs

3.1 National Technological Strategic Planning

The Republic of Singapore is an excellent example of technological strategic planning on a national level. Singapore's technological development strategies have been systematic and purposefully formulated to overcome the country's inherent disadvantages (Yeo 1995). However, current Singapore R&D spending and available personnel are insufficient to overcome these deficiencies. Singapore is considering a strategic division of its R&D efforts to maximize the return from the current level of technological development funding and other technology acquisitions. This is also being done to achieve high utilization of the available pool of technology personnel. Each sector would carry out different technological tasks. Figures 2.6 and 2.7 present the various types of technological tasks and the proposed R&D distribution matrix.

Type of Technological Tasks	State of Technological Applications	Problem-solving Capability Required
Type I With High Certainty	• Apply basic or proven technology already commonly available and in use. • Use basic and well established manufacturing processes and facilities commonly available. • Mainly basic development work in process improvement or basic product design, with no or little research.	• Solving only routine and well structured problems. • Adopting well developed methodology and procedures. • High predictability.
Type II With Controllable Uncertainty	• Adopt known technology that has not been embodied into existing products and processes in local industry. • Mainly new product or process development using proven technology.	• Solving broadly structured problems. • Have broad knowledge of advanced technology and well informed of its potential • Moderate predictability of outcome.
Type III With Many Uncertainty and Ambiguity	• Develop and adopt new and more advanced technology to advanced products or processes not already available. • Advanced product or process development involving sustained research.	• Solving more ambiguous and ill-structured problems. • Have incomplete knowledge of new technology. • Low predictability of outcomes.

Fig. 2.6   Types of Technological Tasks (Source: Yeo 1995, © 1995 IEEE. Reprinted by permission.)

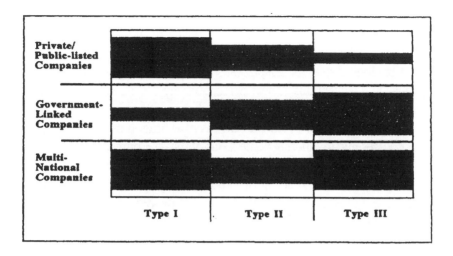

Fig. 2.7 R&D Distribution Matrix Proposed By Singapore (Source: Yeo 1995. ©
1995 IEEE. Reprinted by permission.)

3.2 Corporate Technological Strategic Planning

Western and far eastern enterprises have different approaches to technological
strategic planning.

Western Concepts: Many technology managers in large western
enterprises consider strategy in terms of (Hamel and Prahalad 1995):

- Concept of fit to their enterprise's markets and overall objectives.
- Relationship between the enterprise and its competitive
  environment
- Allocation of resources among competing investment
  opportunities
- Long-term perspective in which *patient money* figures
  prominently.

This is the focus used by the majority of western enterprises in
formulating their technological strategic plans. Some eastern technology
enterprises, primarily those with collaborative arrangements with western
counterparts, also use this approach. The strategic planning phase of
technology is when the innovation within an enterprise becomes legitimized
by key decision makers and further resources are allocated for the
innovation's implementation. It is at this phase the enterprise's need evolves
from (Ramamurthy 1995):

- Compelling enterprise needs (need-pull), or
- Potential technological opportunities (technology-push), or
- Potential competitive needs (competitive-push).

Technological strategic adaptation suggests that enterprises are influenced by managerial perceptions. These perceptions are of a number of internal and external factors. This results in varying choices of appropriate domains for the enterprise operation. Another result is modification of internal structures and processes due to differing perceptions of the environment.

Technological strategic planning systems play critical roles in aiding the enterprise to understand its environment, reduce uncertainty, provide for proactive and rapid response to problems and opportunities. The important aspects of a technological strategic planning system include:

- Scope and comprehensiveness
- Flexibility
- Adaptability
- Accommodation of diverse stakeholders' interests.

Research studies have shown that managerial perceptions are important in identifying needs. Managerial perceptions are also important in motivating and shaping enterprise strategic responses (Ramamurthy 1995).

In a corporate environment, the technological strategic planning process is primarily associated with need-pull and technology-push forces. The important factors of the strategic planning processes within a corporate environment include:

- Conscious strategic long term planning which is a deliberate rationalization of strategies with key, timely and relevant informational inputs and underlying assumptions.
- Depth of analysis of inputs; however, superficial analysis will lead to erroneous results.
- Participation in and support of the planning process by key managers and members of the leadership team.
- Flexible and adaptive technological strategic plan which is capable of being responsive to environmental changes.

Far Eastern Perspective: Japan's success at technological product development is well documented. Japanese enterprises are able to develop new technological products in shorter time periods than similar U.S. enterprises (Fink 1993). Research has shown that Japanese enterprises tend to base their strategic planning process on:

- Multifunctional problem solving
- Close relationships with suppliers and customers
- Incremental improvement
- Learning.

Multifunctional Problem Solving: The development strategies of Japanese enterprises and many of the far eastern technology firms use multifunctional problem solving. These organizations implement technological strategy as an integrated part of the enterprise's business strategy and they integrate the multidisciplinary activities (see Chapter Four - *Generation of Technology*).

Close Customer-Supplier Relationships: Far eastern technological enterprises usually have much closer relationships with customers and suppliers that U.S. organizations. Japanese enterprises receive a larger percentage of their technological R&D concepts from their customers than U.S. firms (Fink 1993). Many Japanese and far eastern enterprises are more vertically and horizontally integrated than similar U.S. enterprises and this results in more co-ordination between dissimilar enterprises. The close relationship between suppliers, enterprises, and customers form a system that can develop an integrated technological strategy for implementation can remove many of the uncertainties in conventional western strategic planning systems.

Incremental Improvement: Many U.S. technology firms employ a strategy that concentrates on radical rather than incremental improvement. Technological products that incrementally build on previous products and markets are more likely to succeed than radical departures in technologies and/or markets.

Japanese and far eastern technology companies spend a significantly larger percentage of their R&D budget on improvement of existing products and processes than U.S. enterprises. Incremental improvements to existing technologies can prevent the implementation of strategies that take advantage of *technological discontinuities.* The research on technological discontinuities implies that a strategy of incremental improvement is only applicable in certain situations. The development of a major breakthrough can offer an enterprise substantial market advantages. A case in point is the development of the Java™ by Sun Microsystems, which caused a major re-evaluation of PC connectivity.

Learning: A learning enterprise is defined as an organization skilled at creating, acquiring and transferring knowledge, and modifying its behavior to reflect new knowledge and insights. Learning for many Japanese and far eastern technological enterprises is an important strategic process and is included in the development of their strategic plans. The incorporation of learning in the total technological strategic planning process will enable an enterprise to change, improve quality, and to develop shorter cycle time and unit costs (Woon 1995).

3.3   Entrepreneurial Strategic Planning

Research indicates that a comprehensive analytical approach to technological strategic planning may not be suitable to new entrepreneurial enterprises (Bhide 1994). These new enterprises generally lack the time and resources to:

- Interview a representative cross section of potential customers.
- Analyze substitutes.
- Reconstruct competitors' cost structure.
- Project alternative technology scenarios.

According to Bhide, many successful entrepreneurs spend little time researching and analyzing, and those who do often have to scrap their strategies and start over. According to a number of studies, successful entrepreneurs do not blindly take risks. Successful entrepreneurs use a quick and low cost approach that represents a middle ground between planning overkill and no planning at all. The successful entrepreneur generally follows these guidelines in developing strategy:

- Screen opportunities quickly to weed out unpromising ventures.
- Analyze ideas parsimoniously by focusing on a few important issues.
- Integrate action and analysis by not waiting for all the answers and being ready to change course.

3.3.1   Screening Opportunities

Entrepreneurs generate many ideas. They also learn to quickly discard those that appear to have low potential or probability of success. This process requires judgment and reflection. Some entrepreneurs like Walt Disney fail many times before finally succeeding. This learning process coupled with an internal belief system and stochastic resonance may be the secret of many individual entrepreneurial successes. Successful entrepreneurial enterprises don't need a technological edge. Microsoft's Bill Gates built a multibillion-dollar technological enterprise without a breakthrough product or technology.

Successful entrepreneurs screen potential ventures or technologies for their risks and rewards compared to other alternatives. The factors that are usually considered include:

> *Capital Requirements* -- favor ventures and technologies that are not capital intensive.
>
> *Profit Margins* -- sufficient to sustain rapid growth with internally generated funds.
>
> *Margin for Error* -- ventures and technologies with simple operations and low fixed costs that are less likely to face cash flow problems due to technical delays, cost over-runs and slow build-up of sales or operation.
>
> *Sufficient Reward* -- to compensate for exclusive commitment to this particular technology or venture.

3.3.2 Parsimonious Analysis

Successful entrepreneurs minimize time and resources devoted to researching ideas. By developing analytical priorities, entrepreneurs recognize that some critical uncertainties cannot be resolved through more research. The frugal entrepreneur does not concentrate on research that cannot be acted upon. The successful entrepreneur concentrates on understanding what must go right and anticipating *show-stoppers*. Analyzing the rewards-risk structure of a potential technology or venture strategy is very important to the entrepreneurial enterprise.

In technology, life cycles are relatively short; thus technologies that provide potentially high initial returns are favored over growing returns that may not fully materialize, so that the entrepreneur and investors may fully recover their investments. Successful entrepreneurs usually devote more attention to operational analysis and planning than strategic planning.

3.3.3 Integrated Action and Analysis

Entrepreneurs do not usually know all the answers before they act. Generally, on many occasions, entrepreneurs cannot easily separate action and analysis. For an entrepreneur, acting before an opportunity is fully analyzed has many benefits. These benefits include:

- Doing something concrete builds individual and group confidence.
- Key employees and investors will follow the entrepreneur who is committed to action.
- Early action can generate more robust, better informed strategies.

The entrepreneurial approach that integrates action and analysis has the following characteristics:

- Handling analytical tasks in stages - Stepwise approach, doing sufficient research to justify next action.
- Plugging holes quickly -- As problems or risk materialize, entrepreneurs seek solutions.
- Evangelical investigation -- Entrepreneurs often blur the line between research and selling. From the beginning, entrepreneurs do not just seek opinions and information, but look for commitment from other individuals.
- Smart arrogance -- An entrepreneur's willingness to act on sketchy plans and inconclusive data is often sustained by an almost arrogant self-confidence.

According to existing research, astute successful entrepreneurs do analyze and strategize extensively. However, they realize new technologies or ventures cannot be launched with every detail planned in advance. Successful entrepreneurs play with and explore ideas, letting their strategies evolve through a seamless process of venturing, analysis and action.

REFERENCES

Betz, F. (1993). *Strategic Technology Management*, McGraw Hill, Inc., New York, NY

Bhide, A. (1994). "How Entrepreneurs Craft Strategies That Work." *IEEE Engineering Management Review*, 22(Winter 1994), 52-59.

Çambel, A. B. (1993). *Applied Chaos Theory: A Paradigm for Complexity*, Academic Press, Inc., Boston, MA.

Fink, J. L. (1993). "Japanese Product-Development Strategies: A Summary and Proposition About Their Implementation." *IEEE Transactions in Engineering Management*, 40(August 1993), 224 - 236.

Hamel, G., and Prahalad, C. K. (1995). "Strategy as Stretch and Leverage." *IEEE Engineering Management Review*, 23(Spring 1995), 2 -7.

Henderson, B. D. (1971). "The Growth Share Matrix of the Production Portfolio." , The Boston Consulting Group, Boston, MA.

Kahn, H., and Wiener, A. (1967). *The Year 2000*, Macmillan, New York, NY.

Martino, J. P. (1993). *Technological Forecasting for Decision Making*, McGraw-Hill, Inc., New York, N.Y.

Mintzberg, H., and Quinn, J. B. (1991). *The Strategy Process: Concepts, Contexts, Cases*, Prentice-Hall, Inc., Englewood Cliffs, NJ.

Pearce, J. A., II , and Robinson , R. B., Jr. (1991). *Strategic Management: Formulation. Implementation, and Control*, Richard D. Irwin, Inc., Homewood, IL.

Porter, M. E. (1980). *Competitive Strategy: Techniques for Analyzing Industries and Competitors*, The Free Press, New York, NY.

Porter, M. E. (1985). *Competitive Advantage*, The Free Press, New York, NY.

Pyke, D. (1973). "Mapping - A System Concept for Displaying Alternatives." A Guide to Practical Technological Forecasting, B. A. Schoedman, ed., Prentice-Hall, Inc., Englewood Cliffs, NJ.

Ramamurthy, K. (1995). "The Influence of Planning on Implementation Success of Advanced Manufacturing Technologies." *IEEE Transactions in Engineering Management*, 42(February 1995), 62 -73.

Shoeffler, S., Buzzel, R., and Heany, D. (1974). "Impact of Strategic Planning on Profit Performance." Harvard Business Review.

SRI. (1963). "A Framework for Business Planning." *162*, Stanford Research Institute, Stanford, CA.

TBCG. (1972). "Perspective on Experience." , The Boston Consulting Group, Inc., Boston, MA.

Thamhain, H. J. (1992). *Engineering Management: Managing Effectively in Technology-Based Organizations*, John Wiley & Sons, New York, NY.

Woon, H. Y., Clement. "Enhancing Product Development: A Case Study." *IEEE Annual International Engineering Management*, Singapore, 92-97.

Yeo, K. T. "Managing Technology Dynamics - A Framework for National R&D Planning." *IEEE Annual International Engineering Management*, Singapore, 327-330.

DISCUSSION QUESTIONS

1. Discuss the similarities and differences in the strategy process for:
   - technology based enterprise
   - commercial enterprise
   - governmental enterprise

2. Prepare a one to two page technological vision, objective and mission statement for an organization that you are familiar with.

3. Choose a technology and prepare a scenario that could be helpful in developing a technological strategy for that technology.

4. Comment upon the differences between the technological strategic process for:
   - European style technology enterprise
   - U.S. style technology enterprise

- Far Eastern style technology enterprise

Why are the approaches taken by these various cultures similar to or different from each other?

5. Comment on how the following features can be used to increase the flexibility of a technological strategic plan:

   - Inclusion of multiple options at future decision points
   - Inclusion of steps to collect additional information before making future decisions
   - Inclusion of intermediate or lesser goals short of completion of the full technological strategic plan

6. Prepare a detailed outline for a Technological Strategic Plan for a selected technology of your choice.

CHAPTER 3

Technological Forecasting

1.   INTRODUCTION

Jules Verne[1], Bellamy[2] and other writers prior to World War II were forecasters of technology. Their forecasts were in terms of fiction, i.e., the realm of fantasy. Technological forecasting today is not a tool of fiction, it is a management tool used for the prediction and estimation of feasible or desirable parameters in future technologies (Edosomwan 1989, p. 96). The objective of technological forecasting is to bring information to the technology management process that attempts to reduce some of the uncertainty about future developments. The management of technology involves an understanding of potential future impact to objectives and missions of the enterprise. However, the future is uncertain; the function of technological forecasting is to reduce that uncertainty where possible or at least understand the possible impacts resulting from stochastic changes in the enterprise's environment.

During the last fifty years various analytical concepts for aiding the technology manager in evaluating the future have been presented. No analytical techniques can remove the uncertainties in the future and give the decision maker a glimpse of a certain future.

Managers of technology are greatly concerned about properly utilizing existing technology and future technological advances. The ability of an enterprise to forecast technological trends and needs that are consistent with the strengths and desires of the organization is crucial to the development of

---

[1] **Verne, Jules,** 1828-1905, a French writer who is considered the founder of modern science fiction. His novels include *Journey to the Center of the Earth* (1864) and *Around the World in Eighty Days* (1873).

[2] See page 61

an effective technological strategy (see Chapter Two - *Technological Strategy*).

1.1   Meaning of Technological Forecasting

Technological forecasting can be defined as the prediction of the future characteristics of useful machines, procedures or techniques. According to Bright (Bright 1968), technological forecasting is a system of logical analysis that leads to quantitative conclusions or a limited range of possibilities about technological and associated economic attributes.

Technological forecasting is related to the tasks which emerge in connection with medium and long-range planning. These forecasts are basically concerned with the rates of technological progress. Jantsch, who conducted some of the initial pioneering studies, defines technological forecasting as (Jantsch 1967, p. 23):

> *"...the probabilistic assessment of future technology transfer, which ... denotes the entire range of vertical and horizontal transfer processes that constitute the advancement of technology and the effectuation of impact in technological (economic, social, military, political, etc.) terms."*

Technological forecasting is not a new approach to predicting the future. In the medieval and renaissance periods, the future was heaven and hell (Gilfillan 1968). According to Gilfillan, the first general predictor of the future, although not a technological forecast, was d'Argenser, a marquis and foreign minister of France who wrote these forecasts from 1729 to 1752. H.G. Wells[3] was one of the early predictors of technology in this century. In 1901, Wells clearly foresaw the characteristics of the automobile, although motor vehicles were rare oddities (Wells 1901).

Technological forecasting emerged as a serious discipline during and after World War II, when objectives, needs, and desires were introduced as

---

[3] **Wells, H (Herbert) G. (George)**, 1866–1946, an English author and social thinker. Having taught biology, he wrote fantastic and pseudoscientific novels like *The Time Machine* (1895) and *The War of the Worlds* (1898). He turned to realism in *Kipps* (1905) and *Tono-Bungay* (1909), and finally to increasing pessimism in *The Shape of Things to Come* (1933).

normative elements in forecasting and constraints were recognized and taken into account. In the 1940s, technology forecasting began, due to the conditions which then existed including a realization that technology had a responsibility towards society. It was also recognized that technology had substantial economic potential. Many foresaw the ultimate technological potential of developments. However, there was also a growing awareness of constraints, for example in resources, i.e., natural, enterprise, etc. Finally, technology was seen as a means to hedge against threats from potential enemies and nature.

In the latter part of the twentieth century, formal technological forecasting emerged as a discipline which attempted to define the probable future capabilities of science and technology and provide planning information to assist technology developments into efficient paths or trajectories. The explosive growth of the United States investment in science and technology, due to the actual threats during World War II and perceived threats after the War by the former Soviet Union, created a requirement for accurate methods of forecasting the probable results of increasing R&D expenditures. It has been stated by Cole (Cole 1965, p. 22):

"We predict the future because we must in order to live."

Prehoda (Prehoda 1967, p. 11-16) has defined technological forecasting as the description or prediction of a foreseeable invention, specific scientific refinement, or likely scientific discovery that promises to serve some useful function or have a significant impact, either positive or negative. At the various technological planning levels forecasting has a different function (Edosomwan 1989, p. 97).

Policy Planning: At this level, technological forecasting provides clarification of scientific technological elements that determine future developmental boundary conditions. Policies can be impacted by various changes in technological developments.

Strategic Planning: The development of useful strategic plans requires a recognition and comparative evaluation of alternative options (see Chapter Two - *Technological Strategy*).

Operational/Tactical Planning: Probabilistic assessment of future technology which technology forecasting can provide offers the technology manager the ability to anticipate potential positive and negative impacts.

Marketing/Corporate Profit Planning: Another level of technological forecasting is to provide a clarification of scientific technology needed to expand market share.   As this chapter will demonstrate, technological forecasting is an excellent tool to forecast market penetration and the impact of the entrance of a new competitor or competing technology.

2.   EXPLORATORY AND NORMATIVE TECHNOLOGICAL FORECASTING

Jantsch's ground-breaking study of technological forecasting set a firm philosophical and analytical ground for technological forecasting.   According to Jantsch, there are two forms of technological forecasting (Jantsch 1967, p. 30), *Exploratory* and *Normative*. An exploratory forecast starts with past and present conditions and projects these to estimate future conditions, while a normative forecast starts with future needs and identifies the technological performance necessary to meet these required needs.

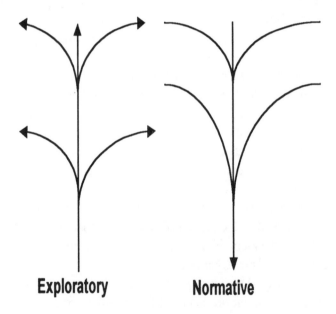

Exploratory   **Normative**

Fig. 3.1 Exploratory and Normative Technological Forecasting (*Source*: Jantsch 1967, p. 30, © 1967 OECD. Reprinted by permission of OECD.)

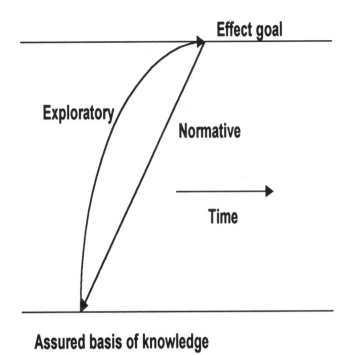

Fig. 3.2 Normative Forecasting Time-Frame (*Source*: Jantsch 1967, p. 31, © 1967 OECD. Reprinted by permission of OECD.)

Basically these types of forecasts are a fundamental polarity of action and reaction (see Figure 3.1). It is important that the interaction of the opportunity-oriented (exploratory) and mission-oriented (normative) forecasting be stated correctly. The basic form of interaction between the two forms of technological forecasting is an interactive *matching* - this is the most difficult aspect of forecasting technological developments (see Figure 3.2). Placing a normative forecast in a correct time-frame is difficult since desires and developments usually do not match. The exploratory aspect of forecasting (the type of forecast usually found in the technical literature) encounters less difficulty in developing an end-point on the basis of a time-span estimate. The objectives and requirements of a normative forecast are usually societal goals based on the assumption that future goals are like present goals. Table 3.1 illustrates the basic difference between normative and exploratory technological forecasting.

84

Table 3.1

Comparison of Exploratory and Normative Technological Forecasting

Level	Exploratory (Opportunity-Oriented) push-approach	Normative (Mission-Oriented) pull-approach
8 Society	Impact on society	Social goals
7 Social Systems	Impact on national, economy, defense, health programs, etc.	National objectives
6 Environment	Consequences for structuring of industry leadership of innovating enterprises.	Missions
5 Applications	Technological economic and social acceptance, measure of success.	Tasks
4 Functional Technological Systems	Description of systems and detailed performance characteristics, development time and effort, production costs.	Relevance of systems to tasks, technological feasibility, cost/effectiveness.
3 Elementary Technology	Functional capabilities, technological parameters	Relevance and feasibility, development gaps.
2 Technological Resources	Basic technological potential	Technological potentials and limitations, required fundamental technological research.
1 Scientific Resources	Trends in scientific principles and theories, unapplied knowledge, applicability to technological progress.	Absolute (natural) potentials and limitations, required fundamental scientific research.

(*Source:* Jantsch 1967, p. 32-33, © 1967 OECD. Reprinted by permission of OECD.)

2.1   Normative Technological Forecasting

When the forecast is "needs oriented" it is termed *normative*. Normative technological forecasting foundation is based upon systems analysis.   A normative forecast starts with future needs and identifies the technological performance required to meet those needs.   The normative approach forecasts the capabilities that will be available on the assumption that needs will be met.   Implicit to the normative approach is that the required performance can be met by a reasonable extension of technological progress (Martino 1993, p. 235).

Normative technological forecasting approaches *self-fulfilling* prophecies. This type of forecasting is meaningful only if the levels to which it is applied are characterized by constraints, and if multiple opportunities exist on these levels than can be exploited under given constraints.

In some instances, the normative approach can be viewed as a planning tool. However, since the technologies which are being planned do not exist, but are means to achieving an end, i.e., the objective, a normative approach for choosing among possible technologies and allocation of limited resources is employed.

The following are normative technological forecasting techniques which can be employed:

- Decision Matrices
  - Horizontal
  - Vertical
- Relevance Trees
- Morphological Analysis
- Network Techniques
- Mission Flow Diagrams
- Other Techniques

2.1.1 Decision Matrices

The principal importance of the matrix for forecasting is the structuring of the thinking and the explicit demand for a forecast of end-uses. At the planning stage, the matrix functions like a simplified PERT[4] scheme and maintains the total systems view.

Horizontal Decision Matrices

As for all normative techniques, horizontal matrices must have a surplus of proposals to analyze. A two or three dimensional matrix is one method for assessing priorities among a number of proposed projects. The most common use of a matrix is for the optimization of resources under given conditions. The matrix is similar to an input/output matrix of scientific and technical efforts in various fields. An example is a matrix for each project,

---

[4] PERT or program evaluation and review techniques

showing tasks and disciplines involved as one dimension and enterprise resources as the other dimension. Projects fitting into the given resource structure are combined into higher-level matrices. Research or market opportunities are frequently used to decide a product mix, chiefly by iteration of the matrix.

Vertical Decision Matrices

Vertical decision matrices can be difficult to construct, since they relate to potential technological developments, actions and resources. One reason for this difficulty is the problem of quantifying relations between different levels. The approach can be characterized as a rectangular prism with three dimensions:

- technology and impact levels
- activities
- applied science discipline or research management spectrum

Figure 3.3 shows a simplified prism.

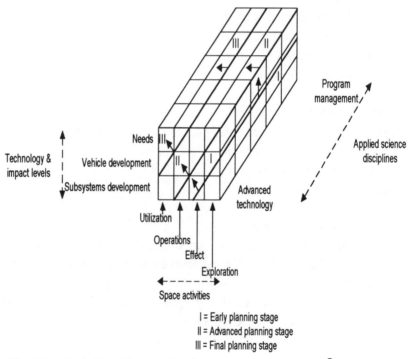

Fig. 3.3 Vertical Decision Matrices (*Source*: Jantsch 1967, p. 213 © 1967 OECD. Reprinted by permission of OECD.)

2.1.2 Relevance Trees

Relevance trees are a tracing method which emphasizes structural relationships, i.e., a chain of causes and effects. Relevance trees graphically depict the linkages between various members of sets of elements, moving from level to level via a relationship (Porter et al. 1991, p. 299). Each of the lower levels involves finer distinctions or subdivisions. Figure 3.4 shows a relevance tree. There is no requirement that each node have the same number of branches.

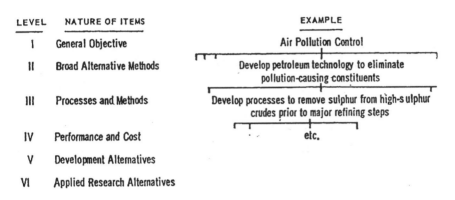

LEVEL	NATURE OF ITEMS	EXAMPLE
I	General Objective	Air Pollution Control
II	Broad Alternative Methods	Develop petroleum technology to eliminate pollution-causing constituents
III	Processes and Methods	Develop processes to remove sulphur from high-sulphur crudes prior to major refining steps
IV	Performance and Cost	etc.
V	Development Alternatives	
VI	Applied Research Alternatives	

Fig. 3.4   Example of Relevance Tree after Jantsch (*Source*: Jantsch 1967, p. 228,. © 1967 OECD. Reprinted by permission of OECD.)

Qualitative relevance trees are employed to aid decision-making, and are also known as *decision trees*. However, qualitative relevance trees can be very useful in quantitative analysis. A relevance tree has the following characteristics:

- Branches depending from a node must be a closed set, i.e., an exhaustive listing of all possibilities.
- Branches depending from a node must be mutually exclusive, i.e., no overlapping.
- Branches must be viewed as goals and subgoals, i.e., each goal is satisfied by the satisfaction of all nodes below it.
- Branches will be either problems or solutions:
- Solution tree - "and" and "or" nodes
- Problem tree - only "and" nodes

Relevance trees can be used to identify problems and solutions by determining the performance requirements of specific technologies. This type of tracing method can be used to determine the relative importance of efforts to increase technological performance.

The matrix for analyzing each level of the relevance tree has the form shown in Table 3.2

Table 3.2
Relevance Tree Matrix

Criteria	Weights of Criteria	Items on Level I						
		a	b	c	....	j	...	n
a	q_a	S_a^a	S_b^a	S_c^a	...	S_j^a	...	S_n^a
b	q_b	S_a^b	S_b^b	S_c^b	...	S_j^b	...	S_n^b
g	q_g	S_a^g	S_b^g	S_c^g	...	S_j^g	...	S_n^g
...	...	...	...	...	...	...	...	...
c	q_c	S_a^c	S_b^c	S_c^c	...	S_j^c	...	S_n^c
...	...	...	...	...	...	...	....	...
n	q_n	S_a^n	S_b^n	S_c^n	...	S_j^n	....	S_n^n
		r_i^a	r_i^b	r_i^c	...	r_i^j	...	r_i^n

S_j^c = Significance number (how significant is the contribution of issue j to criterion c)

r_i^j = Relevance number of item j on level i

The criteria c, the weights of the criteria q_c and the significance number S_j^c are estimated based on a qualitative scenario (see Chapter Two - *Technological Strategy*)). Two normative conditions to assure homogeneity are

$$\sum_{\chi=\alpha}^{v} q_\chi \equiv 1$$

and

$$\sum_{j=a}^{n} S_j^\chi \equiv 1$$

The relevance number is then defined as

$$r_i^j \equiv \sum_{\chi=\alpha}^{v} q_\chi S_j^\chi$$

and

$$\sum_{j=a}^{n} r_i^j \equiv 1$$

Figure 3.5 and Table 3.3 are an example of relevance tree analysis.

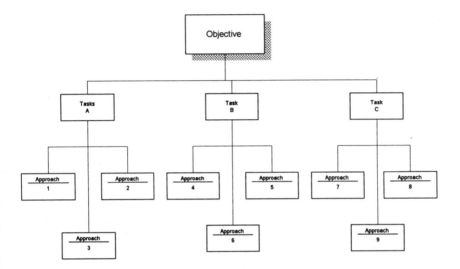

Fig. 3.5 General Relevance Tree

Table 3.3
Task Level Relevance

Criteria	Weights	Tasks		
		A	**B**	**C**
Cost	0.6	0.3	0.6	0.1
Timing	0.3	0.1	0.6	0.3
Future Products	0.1	0.1	0.4	0.5
	Relevance	**0.22**	**0.58**	**0.20**

The total relevance figure of a particular issue on any level is obtained by multiplying upwards to the top of the tree, or down from the top to the level of the issue in question. The use of relevance numbers increases the ability of the approach for identifying needed changes in technology.

2.1.3 Morphological Analysis

A system for breaking a problem down into parallel parts as distinguished from the hierarchical breakdown of the relevance tree is known as morphological analysis (Martino 1993, p. 239). This approach was invented in 1942 by Fritz Zwicky, who employed it for structural analysis of problems (Jantsch 1967, p. 110). In morphological analysis, a fractionalization approach is used to choose parameters of importance to a concept, the alternate possibilities for each are defined, and a checklist is created by making an exhaustive list of all the possible combinations (Porter et al. 1991, p. 105). For example, if a problem can be divided into four independent parameters, and two alternate possibilities for each parameter, then there are a total of sixteen possible combinations. Each combination is a combination of possibilities for the parameters. Some of the combinations will not be feasible. The possibilities can exceed analytical capabilities. Table 3.4 shows an example of the morphological approach for an electric vehicle.

Table 3.4
Fractionalized Components for Morphological Analysis
Electric Vehicle

Power Source	Drive Train	Guidance Mechanism
Primary Battery	Direct Motor	Driver
Fuel Cell	Indirect Motor	Towed
Secondary Battery		Guided Path
Third Rail		Satellite Guidance
Induction		Collision Avoidance
		None

A morphological analysis of Table 3.4 could result in sixty potential systems for analysis. In general, morphological analysis and relevance trees are suited to different types of problems (Martino 1993, p. 239-241). If a system is modeled by both relevance tree and morphological analysis, the elements of the morphological analysis will have a one-to-one correspondence to major connected sections of the relevance tree, and the branches appearing at the bottom level will be the components of the corresponding elements.

2.1.4 Network Techniques

Like the Critical Path Method ("CPM"), the normative technique of a technological network is based on *flow charts*. These flow charts show all branches of technological development. Analysis of these charts results in a selection of an *optimum path* between the first and last stages, where criteria for an optimum may be lowest cost, shortest time, etc. Flow charts, when combined with probabilistic analysis of uncertain data and input relationships, are similar to the noted Program Evaluation and Review Technique ("PERT"). The development of a new technology, requiring large expenditure of resources, cannot be optimized with respect to a single node of its future environment, but must consider the stochastic resonance of the total system. Figure 3.6 is an example of this approach. An important aspect of the network approach is the identification of critical paths (disciplinary areas, functional subsystems, etc.).

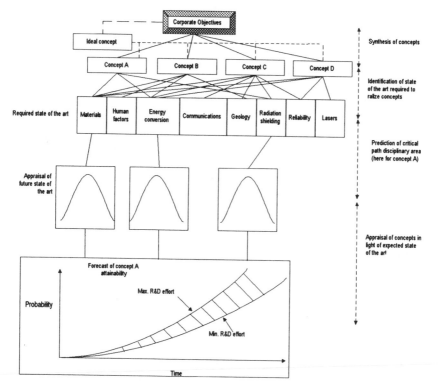

Fig. 3.6 Network Technological Forecasting Flow Chart (*Source*: Jantsch 1967, p. 234, © 1967 OECD. Reprinted by permission of OECD.)

2.1.5 Mission Flow Diagrams

Originally developed as a means of analyzing military missions, mission flow diagrams can be used to analyze any sequential process (Martino 1993, p. 241-242). In this approach, all the alternative routes, or sequences by which some task can be accomplished, are mapped. A mission diagram is prepared (see Figure 3.7), then the problem associated with each route is determined. The cost associated with each route is also determined. The performance requirements for the technologies are derived using numerical weights placed on the alternative paths. This numerical approach is similar to the use of relevance numbers in relevance trees. The results from this analysis are used as a normative forecast.

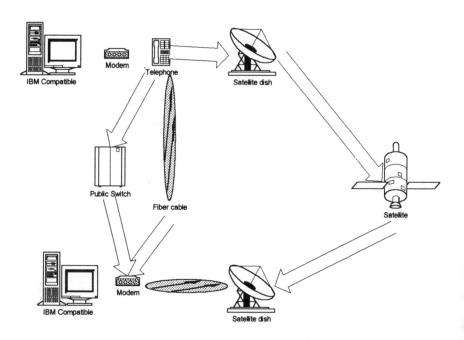

Fig. 3.7 Mission Flow Diagram

An advantage of using mission flow diagrams is in allowing the user to identify problems more readily. Another advantage of the mission flow approach is that it derives the performance required of some technology to overcome the discovered problems. Like both the other normative approaches, numerical weights can be placed on the alternative paths of a mission flow diagram. These values can be used to determine the relative importance of the technological solution to each of the bottlenecks or difficulties for the paths.

2.1.6 Other Techniques

Other normative technological forecasting techniques, while not new, have not been significantly applied to technological forecasting, even though they may offer certain advantages. These techniques include game and complexity theory.

Game theory uses a mathematical approach to select optimum strategies. Technology options would be used either in pure or mixed strategies. The strategies could be applied with nature or an actual competitor as the opponent. The role of chance events, matrix representation of pay-offs, etc. is also employed. The use of game theory can yield simple optimum strategies such as maximization of profit, but optimized mixed technological developmental strategies in circumstances where any single strategy would involve vulnerability to potential negative possibilities.

Complexity theory is a potential approach which has not yet been utilized in technological forecasting. It may be possible to use these new concepts concerning complex systems to offer a means to view the technological developmental process in a new light. This includes combining the concept of technological vectors with complexity theory to arrive at a condition of a *strange technological attractor*, i.e., a technological position from which it would be difficult for an enterprise to easily change.

2.2 Exploratory Technological Forecasting

By using measurable technical data, exploratory technological forecasting has an advantage over normative forecasting. Technical data, according to Jantsch, can be grouped into three sets (Jantsch 1967, p. 143): functional capabilities, technical parameters, and scientific and technical findings. One set consists of the functional capabilities, independent of any specific technology. The technical data can also be grouped by technical parameters. The final set consists of the scientific and technical findings for which the relationship to functional capabilities has not yet been established. These data sets, representing past history, lend themselves to extrapolation over time by using either trend models or a Delphi approach.

2.2.1 Trend Models

Time-series is the principal quantitative approach used for technological forecasting in the early pre-innovation stages of technological development. Extrapolation uses the past to anticipate the future. The underlying trend is that technical attributes often advance in a relatively orderly and predictable manner (Porter et al. 1991, p. 139). The general technological vector for any technology results in orderly technological change, since a complex mix of influences moderates possible discontinuities. All technologies reach a stage where the basic underlying physical laws moderate and limit the advance.

Trend analysis methods can provide useful forecasts when the underlying structures within the environment remain relatively constant over the time horizon of the forecast (Porter et al. 1991, p. 169). Trend analysis can be divided between trend extrapolation and growth curves

Trend Extrapolation

The most common long-term trend is exponential growth. The upper limit is set by the basic physical laws that govern the phenomena utilized by the particular technology. When a technology reaches its limits, a new technology may arise that utilizes a different set of physical or chemical phenomena (Martino 1993, p. 79). The new technology may be subject to its own ultimate limit, but this can be higher than the prior technology. This is considered a *naive model* since the model does not explicitly account for the effects of the surrounding environment on technology.

Some technologies appear to follow an exponential growth pattern for a portion of the growth cycle. Exponential growth is growth by a constant percentage per unit time, i.e., growth is proportional to the value already reached and is given by:

$$\frac{dy}{dt} \equiv ky$$

where k = constant of proportionality; the solution of this equation results in:

$$y = y_o e^{kt}$$

Exponential trends are plotted on a semilogarithmic scale (see Figure 3.8) and represented by the equation:

$$\ln y = Y = \ln y_o + k\,t$$

where

y_o = initial value
k = growth curve
t = time

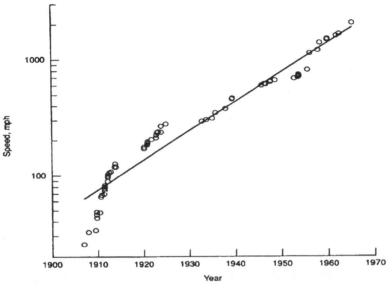

Fig. 3.8 Exponential Trend of the Growth of Flight Speeds Over Time (*Source*: Martino 1993, p. 81. Reprinted with permission of The McGraw-Hill Companies.)

Various mathematical methods can be used to develop similar models, including linear regression. An important extension to simple regression analysis is to consider, simultaneously, the causal influence of more than a single explanatory (independent) variable (Porter et al. 1991, p. 165).

According to Porter et al, there are a number of steps in developing a trend analysis. The steps include choosing the technological variables which can influence the forecast. As previously discussed, this is not a trivial matter since the forecaster may overlook a variable which may at first glance not appear important. To reduce this problem, a systems approach is important. By reviewing the potential interaction of environmental variables such as the influence of socio-economic factors an understanding of the underlying factors influencing a forecast can be developed.

The acquisition of the necessary data is the next important step in the process. This may require researching historical records, and arranging interviews and possibly focus groups. In many instances, forecasters truncate this function. Because of the non-linearity of the underlying factors that influence technological development, it is important to obtain as extensive a data set as possible.

Once the variables have been specified and data has been collected, the forecaster's next activity is to identify the proper trend model to use to develop the forecast. Various models are available to use with the collected data (see next section). The forecaster uses the selected model to fit the data. A graphical or an analytical methodology can be used to obtain the constants required to produce the forecast.

The derived model relationships are then employed to perform a sensitivity analysis. This includes developing statistical confidence intervals for the resulting forecasts. Based on this analysis an interpretation of the technological projections is then developed. This analysis of the projections should also include a realistic appraisal of outside factors which may influence the technology forecast.

Trend Models

The prominent trend models available to the forecaster include: S-shaped, learning, exponential and linear growth curves. These models form the basic *tool kit* used by technological forecasters. Each model has advantages and disadvantages. None of these tools can assure that the forecast is correct due to the stochastic nature of the environment. A major unexpected technological breakthrough or societal change can completely change the outcome. However, these models help explain some of the underlying mechanisms which cause growth and competition between technologies.

S - Shaped Growth Curves

The sigmoidal curves known, also as *S-shaped*, describe many natural phenomena. The basic model for S-shaped curves is based upon the assumption that incremental information depends only on (Jantsch 1967, p. 145):

- Number of investigators
- Recognized upper growth limit

- Communications factor dependent solely on the number of investigators

In the simplest form of the relationship, without a recognized limit and communication factor, the information gain is given by:

$$\frac{dI}{dt} = qN(t) = q\,N_0\,e^{ct}$$

where

I	$=$	information (state of knowledge)
t	$=$	time
q	$=$	average productivity factor per investigator and time unit
$N(t)$	$=$	number of investigators engaged in time t
N_o	$=$	number of investigators engaged at time t = 0
c	$=$	coefficient (slope of curve in logarithmic plot)

Integration of the above equation over time T results in:

$$I = qN_0 \int_0^T e^{ct}dt = q\frac{N_0}{c}\left(e^{cT} - 1\right)$$

Introducing a correction factor for the upper limit ("L") leads to the general S-shaped curve.

$$\frac{dI}{dt} = qN_0\,e^{ct}\left(\frac{L-I}{L}\right)$$

Therefore

$$I = L\left(1 - e^{-\frac{q\,N_0}{cL}e^{ct}}\right)$$

98

Figure 3.9 shows the above equation. Adding a communication factor to account for the link between investigators results in:

$$I = \frac{qN_0^2}{4c}\left(e^{2ct} - 1\right)$$

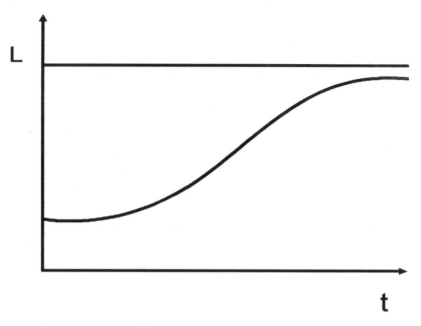

Fig. 3.9 S-Shaped Forecasting Curve

This equation produces the relationship between the general rise of total scientific and technical knowledge. However, this equation does not adequately express typical team technological progress. The equation is presented here merely to show the development of the S-shaped technological curve.

Another S-shaped growth curve approach, which seems to represent technological growth more adequately, is the use of biological competitive models. This form of model is characterized by the work of Pearl[5] (1924/25) and Fisher[6]-Pry[7] (1971) (Porter et al. 1991, p. 176-177). Pearl's work deals

[5] **Pearl, Raymond**
[6] **Fisher, John**
[7] **Pry, Robert**

with the rate of increase of fruit flies within a bottle; rate of increase of yeast cells in an environment; and the rate of cell increase in white rats. In 1971, Fisher-Pry modified Pearl's earlier model and applied it to technological growth (Fisher and Pry 1977). This model is also known as a substitution model because it is applied to forecast the rate at which one technology will replace or substitute for another (Porter et al. 1991, p. 176). The Fisher-Pry model results in

$$f = \frac{Y}{L} = 0.5\left[1 + \tanh a(t - t_0)\right]$$

where

Y = number of units sold
L = upper bound for growth of Y

This equation can be reduced to Pearl's growth equation of:

$$f = \frac{Y}{L} = \frac{1}{\left\{1 + e^{[-b(t-t_0)]}\right\}}$$

Figure 3.10 illustrates the Fisher-Pry substitution curve fits for various products and processes. Another approach to S-shaped curves is that employed by Gompertz[8]. Gompertz and Fisher-Pry models differ in their underlying assumptions. The Fisher-Pry is a growth model while the Gompertz model is a mortality model based on the mortality rate of a population as it grows exponentially as the population ages.

The Gompertz model is appropriate in technological forecasting when technological replacement is driven by technological deterioration, i.e., the technology can no longer meet market or societal demands. The Gompertz model is given by (Martino 1993, p. 63, Porter et al. 1991, p. 183):

$$f = \frac{Y}{L} = e^{-\left(be^{-kt}\right)}$$

---

[8] **Gompertz, Benjamin,** an English actuary and mathematician

Porter et al. shows a comparison for both models for the growth of cable television ("CATV") subscribers in the United States (see Figure 3.11) (Porter et al. 1991, p. 184).

The choice of the type of model to use, i.e., Fisher-Pry or Gompertz, depends upon what technology is being forecast. The Fisher-Pry model is useful when the technological growth is proportional to both the fraction of market penetrated and the fraction remaining and the diffusion of the technology. The diffusion is the given by the number of uses for which the technology has been applied and the number to which it is yet to be applied.

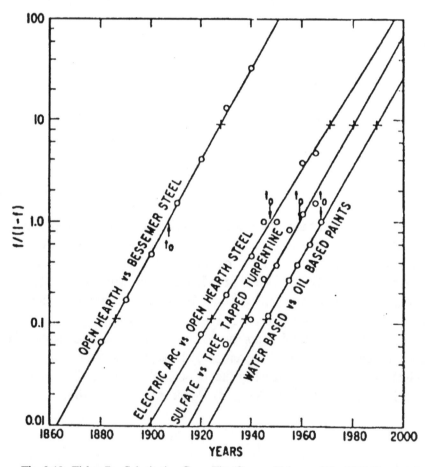

Fig. 3.10 Fisher-Pry Substitution Curve Fits (*Source*: Fisher and Pry 1977. Reprinted by permission of the publisher Copyright 1977 by Elsevier Science Inc.)

The Gompertz model is useful for penetration greater than fifty percent. The rate of penetration depends primarily on the fraction of the remaining market. The model is also appropriate for forecasting market penetration by technologies for which the initial sales do not make subsequent sales easier; penetration is solely a function of time.

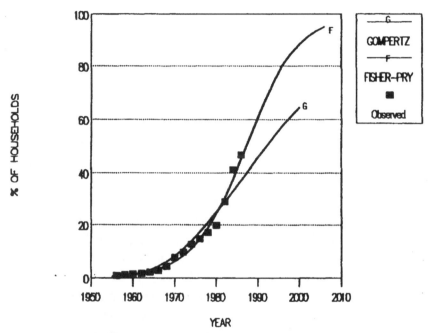

Fig. 3.11 Comparison of Gompertz and Fisher-Pry Models as per Porter et al. (*Source*: Porter et al. 1991, p. 184, © 1991 John Wiley & Sons. Reprinted by permission of John Wile & Sons, Inc.)

Both models depend on selection of an upper bound, i.e., "L". This upper bound must also be estimated. According to Porter et al. (Porter et al. 1991, p. 187), this upper bound should be set by the natural or fundamental limit to the technological process. To determine this upper bound, it is important the preparatory analysis develop an understanding of the technology and the potential market.

It can be shown that many of the trend models are a specific form of the more general equation developed by Lotka-Volterra which was used to model the competitive-resource chain such as the population changes of sharks and the fish upon which they feed (Porter et al. 1991, p. 187).

According to Porter et al., the strong biological basis for the Lotka-Volterra model suggest that the underlying sociotechnical processes lead to the growth and diffusion of technologies (Porter et al. 1991, p. 187).

The Lotka-Volterra equation describing the interaction between two technologies is given by:

$$\frac{dx}{dt} = x \left(\alpha_1 - \beta_1 x - \gamma_1 y \right)$$

$$\frac{dy}{dt} = y \left(\alpha_2 - \beta_2 y - \gamma_2 x \right)$$

This model assumes that technology develops within a larger market system in which technologies may compete, i.e. "sharks" and "fish". This system tends to be open to many forces not directly included in the model, but the competitive nature of the model assumes that it does capture the essential dynamics. The basis of many of the forecasting models can be reduced, under certain conditions, to a form of the Lotka-Volterra model (Porter et al. 1991, p. 191). It must be remembered that under certain initial conditions, this and the other non-linear models can become chaotic in their behavior. Aggressive competition by a major market leader, while usually of initial benefit to the users, can, under certain conditions, lead to the reduction of technological innovation. This occurs when one technology leader captures the dominant market share. This dominance results in forcing any remaining competitors to reduce investment in technological development.

2.2.2 Delphi Approach

The Delphi approach to technological forecasting relies on the knowledge of a committee of experts to extrapolate present technological trends into future technologies. The Delphi approach is intended to gain the advantages of groups of technologists while overcoming their disadvantages (Martino 1993, p. 16-17) . The Delphi forecasting methodology relies upon anonymity since members of the group do not know which members contributed particular statements or opinions. The interaction between the members takes place in a totally anonymous manner through the use of questionnaires. The process also employs iteration with controlled feedback. The interaction takes place through multiple perturbations to the questionnaires. A panel moderator extracts pieces of information relevant to the technological issue under study and presents this to the group. Only the current state of the collected opinions and arguments for and against are returned to the panel for the next iteration. The moderator presents a statistical response that includes the technological forecasting opinions of the entire group.

The guidelines for performing a Delphi forecast include (Martino 1993, p. 24-27):

- Obtain agreement to serve on the panel.
- Explain the Delphi procedure completely.
- Make the questionnaire easy.
- Limit the questions to no more than 25.
- Contradictory forecast can be included in the rounds if generated by the panel.
- The moderator's opinion can be included to help resolve a question by including an argument or fact that both sides of a debate may have overlooked.
- Payment to panelist for their professional efforts.
- Use computer analysis to reduce workload.
- Rapid turnaround between questionnaires.

2.2.3 Scenario Approach

Scenario construction (see Chapter Two - *Technological Strategy*) is useful when time series data, experts or models do not exist. This approach can also be used to help integrate forecasts derived by more analytical approaches such as growth curves. However, scenarios are most useful when no other technique is available. The major problem with scenario construction is that it may produce a more fantasy forecast unless a firm basis in reality is maintained by the forecaster.

3. TECHNOLOGICAL FORECASTING PROBLEMS

Many forecast of future technologies have not necessarily materialized. The majority of the errors of past forecasts have been instances of underestimating possible progress. However, there have also been errors of overestimation. According to Martino (Martino 1993, p. 305-323), there are three major categories of causes of error in technological forecasting: environmental factors, personal factors, and core assumptions used in the forecast development.

3.1 Environmental Factors

There are a number of environmental factors that can impact a technological forecast. The *technological* factor error results from a number of factors. These factors include failure to consider all the stages of innovation, particularly the early stages, ignoring developments in other fields, ignoring the competition between technologies, and assumption of the state and level of technology. An overly optimistic or pessimistic estimate on technology acceptance and growth is an *economic* error factor. A *managerial* factor is the inability of the forecaster to discern the impact of advances in managerial technology.

The changes in the *political* environment can also have significant impact on technological progress. The failure to account for change within *society*, particularly population growth, distribution, financial resources, and special interest groups can introduce errors into the forecast. Changes in *cultural* values may have a significant impact on a technological forecast, since society's needs may change and thus impact the growth of technology. Values of the *intellectual* leadership of society will also cause errors. The growth of the environmental movement and the resulting changes in technology is an example started by **Rachel Carson** (1907–64), a U.S. marine biologist and author who in 1962 published *The Silent Spring*.

Even though at the present time it seems unlikely that any religious group could have a large impact on technological development, there is evidence of a growing *ethical* movement against technology (Johnson 1996). If history has taught us anything, it is that religious movements can arise that can change world views. *Ecological* factors can also interject error into a technological forecast, which is having a growing impact on technological change. The introduction of ISO 14000 will likely cause modifications to the technological manufacturing process offering both opportunities and threats.

3.2 Personal Factors

There are a number of personal factors which can impact technological forecasts. The forecaster, in some instances, may have a *vested interest* in an enterprise or a particular outcome which might be threatened by a particular change. This either purposefully or unconsciously may cause the forecaster to bias the outcome to favor an underlying interest. The forecaster may also have a *narrow focus on particular technology or technological approach*. This happens when a forecaster only looks at a particular technology or technological approach without considering alternatives for accomplishing the same objective. A technological forecaster may have a *commitment to a previous position* and this inflexibility may introduce error into a forecast. A forecaster can also *overcompensate* in the development of the technological forecast. By overcompensating, the forecaster either finds it difficult to present a particular forecast or has a strong desire to see a particular outcome and so distorts the forecast.

Error can be introduced into a forecast by *giving excessive weight to recent evidence*. New evidence, or events, while not fully investigated, may bias a forecaster due to the closeness to the event or evidence. As the study of history has shown, closeness to events can significantly bias results, which also includes *excessive emphasis on recent problems*. Prior technological problems may also bias a forecaster by predisposing the forecaster toward a particular forecast.

In some instances, a forecast may produce a result which would dictate an *unpleasant course of action*. A forecast may appear so unpleasant that the forecaster backs-off from making the forecast, in the hope of avoiding the necessary action - *non-Freudian Oedipus effect*. The forecaster may acquire a *dislike for the source* of the innovation. Innovations arise in many unsuspected places; an example is when Steve Jobs[9] started Apple Computer, the industry ignored the innovation due to the source. Forecasts have to consider a span of time which can be longer than the vision of the forecaster. This causes a systematic *shift from optimism to pessimism* as the time length of the forecast increases. This is difficult to overcome due to the nature of human heuristics.

3.3 Core Assumptions
The underlying assumptions used by the forecaster in constructing the forecast are core assumptions. Once the core assumptions are made,

---

[9] **Jobs, Steven Paul**, 1955–, an American businessman; Working with **Stephen Wozniak**, in his family's garage created the company that produced the Apple computer in 1976. He resigned (1985) and founded the NeXT Computer Company.

selection of the forecasting method may be obvious or is trivial. Therefore, the forecaster should develop these assumptions with great care.

4.   APPLICATION OF TECHNOLOGICAL FORECASTING

Technological forecasting has been used extensively in the last fifty years. The original impetus was for forecasting the technology needed for military purposes. In the mid-1960s, a great deal of interest started to be generated for using technological forecasting in the private sector environment.

4.1   Private Sector Technological Forecasting

The need for developing technological forecasts is growing due to the rapidly changing nature of technology. Many technological developments have been forecasted. Among these developments one which has received a great deal of attention is high definition television ("HDTV").

4.1.1 High Definition Television

The most ambitious form of *Advanced Television* is HDTV. This form of television has been discussed in the United States and Japan as the next generation of television. By using home appliance industry data, it is possible to develop a diffusion model for appliances (Bayus 1994). Figures 3.12 and 3.13 present published forecasts for HDTV. These models indicate the eventual size of the HDTV products market.

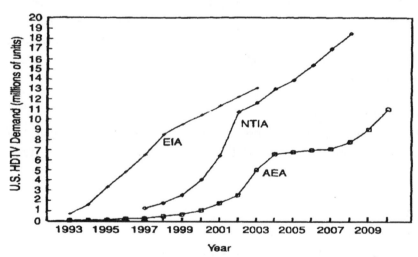

Fig. 3.12   Comparison of HDTV Demand Forecasts (*Source*: Bayus 1994, © 1994 IEEE. Reprinted by permission.)

These models indicate that the forecast by the American Electronics Association ("AEA-1988") over the period 1996-2007, is consistent with prior sales histories of similar home appliances. However, these forecasts do not include the extent that technologies associated with HDTV, e.g., semiconductors, flat-panel displays, depend on establishing a foothold with a consumer product.

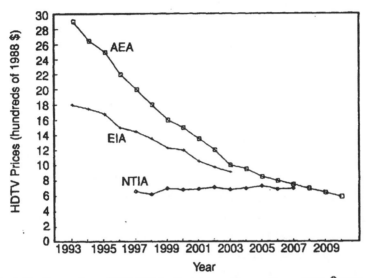

Fig. 3.13 Comparison of HDTV Price Projections (*Source*: Bayus 1994, © 1994 IEEE. Reprinted by permission.)

Telecommunication Technologies

Telecommunications will increase its impact on society during the coming decades. Improved technologies and new market opportunities are two of the principal reasons why the telecommunications industry has been in a state of momentous change since the 1980s. General trends for telecommunications include:

- Liberalization of markets
- Global orientation of goods and services
- Appreciably increasing R&D expenditures
- External growth in the manufacturing sector

A study of the Korean telecommunications industry indicated ten core technologies on which R&D activities should focus from 1992 to 2006 (Suh

108

et al. 1994). While the objectives of this study were primarily for planning purposes, many of the technological areas had not been developed. These technologies were:

- Switching system for B-ISDN
- Synchronous optical transmission systems
- Multimedia terminals
- Network operations and management
- System engineering
- Software
- Satellite communications
- Improved personal communications
- Components for telecommunication systems
- Computers for telecommunications

A normative approach using the Analytical Hierarchy Process ("AHP") was employed for prioritizing technologies, as shown in Figure 3.14. The process was divided into two major steps. The first step is the construction of a six-level hierarchical model of all relevant factors, identifying critical categories at each level and their inter-relationships. Once this is completed, representatives for R&D divisions reviewed the hierarchical structure and devised a priority matrix for each level. Matrices were used to develop a baseline budget level and scope of research activities (see Chapter Six - *Enterprise Structure and Design*).

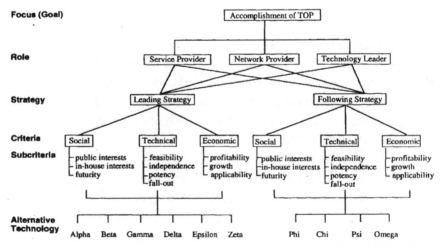

Fig. 3.14 Hierarchical Structure for Analytic Hierarchy Process (*Source*: Suh et al. 1994, © 1994 IEEE. Reprinted by permission.)

4.2 Governmental Sector Technological Forecasting

The U.S. Department of Defense ("DOD") has been the principal pioneering group in using technological forecasting. Agencies like National Aeronautics and Space Administration ("NASA"), Department of Energy ("DOE"), National Oceanic and Atmospheric Administration ("NOAA") and other technology based agencies practice technological forecasting. The following is an example illustrating the approach used by the United States Army in developing their master technological plan (Army 1993).

With the fall of the Soviet Empire, the world has markedly changed from a national security perspective. The collapse of the Soviet Union and the end of the Cold War, which brought about changes in the U.S. defense policies, introduced changes in the U.S. Army's science and technology program. The U.S. Army technological forecasting program relies upon normative and exploratory techniques. However, the majority of the forecasts are normative, i.e. technological-push.

Starting with missions, the U.S. Army determines which DOD technological areas would have impacts on critical science and technology areas. Figure 3.15 shows this matrix.

The U.S. Army has developed normative forecasts, similar to that illustrated for tungsten alloy in Figure 3.16, for other advanced materials (see Figure 3.16). These normative forecasts drive most of the allocation of R&D funding for new technologies. Figure 3.17 illustrates a similar view for automatic target recognition ("ATR") systems for tank warfare.

This type of forecasting can lead to various problems. According to Martino (Martino 1993, p. 243), normative methods have a disadvantage, which they share with all systematic methods of problem solving, namely they tend to impose rigidity on the proposed solutions. This type of forecast difficulty arises in developing numerical criteria, since the values tend to create their own validity.

Many governmental agencies employ normative forecasts. These forecasts are driven by the need to assure funding in future years for current technological developments. The objectives of governmental organizations are established by law and are, therefore, beyond their control. Furthermore, the budgets of these agencies are determined by the legislature, and even the sizes of their workforce are fixed by factors over which they have little control.

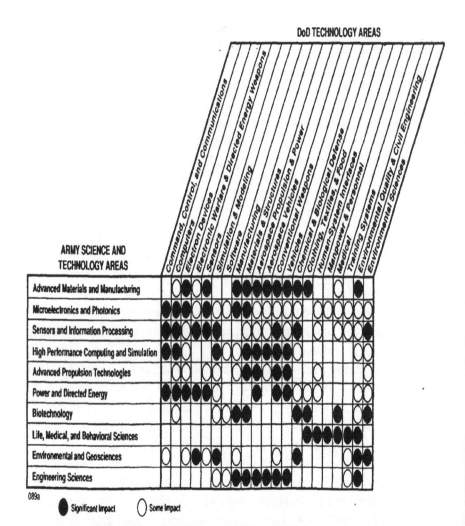

Fig. 3.15 Technological Development for the United States Army Science and Technology Areas (*Source*: Army 1993, p. IV-B-1)

The relationship between the governmental forecaster and the legislature is not direct but proceeds through agency decision makers and senior career executives. This process, coupled with the lack of measurable market factors, has serious problems when normative forecasts are employed, i.e., they can become *self-fulfilling prophecies* (Jantsch 1967, p. 34). In some instances, these forecasts overly optimistic or over-pessimistic depending upon the agency leadership team's disposition due to perceived political constraints.

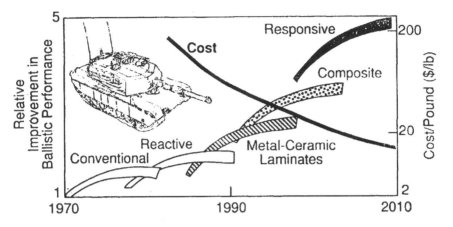

Fig. 3.16 United States Army Advanced Materials Normative Forecast (*Source*: Army 1993, p. IV-C-2)

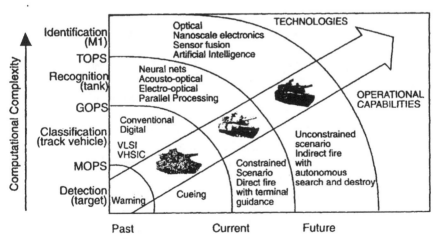

Fig. 3.17 United States Army Normative Forecast for Automatic Target Recognition Technology (*Source*: Army 1993, p. IV-C-2)

Technological forecasting is a difficult process within any enterprise. Within governmental agencies, the technological forecasting process is further compounded by the political environment in which these agencies must operate. Any technological forecaster must strive to develop forecasts that assist the leadership team to form decisions that can assist in meeting the enterprise's objectives.

REFERENCES

Army. (1993). "Fiscal Year 1994: Army Science and Technology Master Plan.", United Sates Army, Washington, DC.

Bayus, B. L. (1994). "High-Definition Television: Assessing Demand Forecasts for a Next Generation Consumer Durable." *IEEE Engineering Management Review*, 22(Fall), 65 - 75.

Bright, J. R. (1968). "The Manager and Technological Forecasting." Technology Forecasting for Industry and Government: Methods and Applications, J. R. Bright, ed., Prentice-Hall, Inc., Englewood Cliffs, NJ, 343-369.

Cole, D. M. (1965). *Beyond Tomorrow: The Next 50 Years in Space*, Amherst Press, Amherst, WI.

Edosomwan, J. A. (1989). *Integrating Innovation and Technology Management*, John Wiley & Sons, New York, NY.

Fisher, J., and Pry, R. (1977). "A simple Substitution Model of Technological Change." *Technological Forecasting and Social Change*, 3, 75 - 88.

Gilfillan, S. C. (1968). "A Sociologist Looks at Technical Prediction." Technological Forecasting for Industry and Government: Methods and Applications, J. R. Bright, ed., Prentice-Hall, Englewood Cliffs, NJ,, 3 -34.

Jantsch, E. (1967). *Technological Forecasting in Perspective*, Organisation for Economic Co-operation and Development, Paris, France.

Johnson, D. (1996). "Technology Is Unwelcome at Gathering of Modern-Day Luddites." The New York Times, New York, NY.

Martino, J. P. (1993). *Technological Forecasting for Decision Making*, McGraw-Hill, Inc., New York, N.Y.

Porter, A. L., Roper, A. T., Mason, T. W., Rossini, F. A., and Banks, J. (1991). *Forecasting and Management of Technology*, John Wiley & Sons, New York, NY.

Prehoda, R. W. (1967). *Designing The Future: The Role of Technology Forecasting*, Chilton Book Company, Philadelphia, PA.

Suh, C.-K., Suh, E.-H., and Back, K.-C. (1994). "Prioritizing Telecommunication Technologies for Long-Range R&D Planning to the Year 2006." *IEEE Transaction in Engineering Management*, 41(August), 264 - 275.

Wells, H. G. (1901). "Anticipation of the Reaction of Mechanical and Scientific Progress Upon Human Life and Thoughts.", London, England.

DISCUSSION QUESTIONS

1. Choose five variables on which an explorative forecast could be made (such as energy use, GNP per capita, etc.). How could a normative forecast be made for the same variables.

2. Chose several technologies, find data and plot the changes over time, compare the plotted data to various models discussed in this unit including Pearl, Gompertz, and Fisher-Pry models. Discuss this analysis in detail.

3. Place yourself in 1974 and based on available data predict what would possibly happen by the year 2000 in the following technologies:
   - Spaceflight
   - Electricity from nuclear power
   - Electric vehicles

4. Prepare an explorative technological forecast for the technology your class group is studying. Each member of the group is to use a different forecast methodology. Comment on your forecast in detail giving both positive and negative aspects.

5. Prepare a normative forecast for the technology your group is studying; each member of the group is to use a different normative forecast methodology. Comment on your forecast in detail giving both positive and negative aspects.

CHAPTER 4

Generation of Technology

1. INTRODUCTION

Generation of technology is the process by which enterprises produce technology. The generation of technology is both an individual and a group activity; however, its management is solely an enterprise activity. This chapter's objective is to present various stages of the generation of technology and its management. Technology must be translated into a useful product, process or service, otherwise it will just remain a technological curiosity, e.g., fusion power or high temperature superconductivity. The progression from initial technological discovery or creativity stage to product, process or service which serves a market is an acquisition process. It is an acquisition process in that the enterprise either acquires the technology through internal development or from external sources.

Technology develops through a process of creativity, invention and innovation. The success of one organization versus the failure of another enterprise has often been traced back to creativity and innovation (Thamhain 1992, p. 5). An organizational environment where people perform creatively and innovatively is a major management responsibility.

Every enterprise is involved in technological change either as an originator, user, or victim of technological invention and innovation (Howard Jr. and Guile 1992, p. 7). The development of a technology process is an evolutionary process in which technology and business structures evolve together.

2. CREATIVITY

The dictionary defines the verb *create* as (Stein 1966):

> "1. to cause to come into being, as something
> unique that would not naturally evolve or that is

*not made by ordinary process....2. to evolve
from one's own thought or imagination, as a
work of art, an invention, etc."*

It is the second form of the definition that we usually associate with creativity. According to Porter et al. there are five key elements of creativity (Porter et al. 1991, p. 100-101). One of the elements is entitled *fluency*, and is the ability to provide ideas in volume. Leonardo da Vinci[1] and Thomas Edison[2] are excellent examples of this element. Each individual developed numerous concepts and inventions which are still having impacts into the twentieth-first century. Both Leonardo da Vinci and Thomas Edison had *flexibility*, i.e., the ability to bend familiar concepts into new shapes or to jump from old concepts to new ones.

Originality is another of the elements of creativity and consists of the unusualness of ideas. Creative individuals also have an *awareness* in that they have imagination to perceive connections and possibilities beyond the obvious. Probably one of the most important elements is *drive* or motivation. Without this element an invention may be conceived but never become embedded within an intended market. Creativity is both individual and a group ability, and it can be stimulated through various techniques.

2.1 Stimulation of Individual Creativity

Porter et al. has identified at least five techniques for the stimulation of individual creativity (Porter et al. 1991, p. 101). Established patterns of thinking tend to cluster into groups that grow larger and larger, eventually becoming dominant patterns. *Lateral thinking* provides a way to restructure and escape from obsolete patterns and to develop patterns which may be more beneficial. This new approach encourages full use of natural pattern-making capacity without hindering creativity, and it generates new paths simply for the sake of finding the range of alternatives.

One approach to developing lateral thinking is by *suspending judgment*. The use of either positive or negative judgment too early in the creative process can be crippling. Suspending judgment, while the creative process is in the early stages, increases the potential for arriving at a creative solution.

---

[1] **Leonardo da Vinci** (1425–1519), Italian artist, scientist.

[2] **Edison, Thomas Alva**, 1847–1931, an American inventor. Edison was a genius in the practical application of scientific principles and one of the most productive inventors of his time.

Dismantling or the *fractionation* of a problem into parts or fractions is another lateral thinking approach. This approach is used to reduce the complexity of the problem. When individual solutions are found, the parts or fractions are assembled in the original problem. *Reversal* can also be employed to develop lateral thinking. In reversal, a problem is turned around, inside-out, upside down or back to front to see what new patterns emerge. This approach is used to find different perspectives by forcing the adoption of a new vantage point.

Another approach to lateral thinking is the use of *metaphors* and *analogies*. Metaphors are words or phrases applied to concepts or objects that they do not literally denote, and they can be used to stimulate lateral thinking. Analogies can also be used to express recognition of similarities between otherwise dissimilar things. These concepts when visualized are much more effective than abstract concepts and should lead to action.

Checklists are an important tool for stimulating individual creativity. In this approach, the manager of technology lists all the attributes of interest such as: adapt, modify, magnify, minify, substitute, rearrange, reverse, combine. According to Porter et al. (Porter et al. 1991, p. 105), checklists must be carefully constructed to allow for exercise of latitude and imagination.

The *morphological analysis* approach, as shown in Chapter Three - *Technological Forecasting-* combines the concepts of fractionation and checklists. This powerful technique is used for stimulating both individual and group creativity. Another approach in lateral thinking is the use of *random words*. Porter et al. believes that this approach often brings about a fresh association of ideas and triggers new concepts or offers a new perspective of familiar ideas.

2.2 Group Creativity

Technologists often work as members of a group. These groups can utilize various techniques to increase creativity. *Brainstorming,* an old concept, has been studied by industrial physiologists since the early 1950s. In the formalized concept, the members of a group are asked to respond to a central problem or theme. Emphasis of the brainstorming approach is on generating a large number of ideas and suspending criticism. The general guidelines for a formalized brainstorming session include ruling out criticism, i.e., no idea is initially unacceptable. The sessions are *free-wheeling* and wild ideas are welcomed. The participants are encouraged to combine ideas and thus expand their concepts.

A number of problems are associated with brainstorming. The process by its nature delays evaluation until all ideas have been gathered. This may cause some participants to lose focus. Dominant individuals of this group may influence other participants and try to monopolize the process. Groups engaged in brainstorming sometimes become so excited that a bandwagon and other consensus phenomena can undermine creativity. During the brainstorming process, it is difficult to have the proper reference material on hand since the ideas are basically stochastic in nature. It is very difficult to keep participants from becoming emotionally involved during a brainstorming session.

Slip writing is another creativity-stimulating process. This technique, which originated with Professor C. Crawford at the University of Southern California, can generate many ideas in a very short span of time (Porter et al. 1991, p. 110). Each group member is given three-by-five inch index cards or slips of paper. The problem is stated and the group is asked to refrain from judgment. The group is then told to write as many ideas as possible within an allotted time. Each idea is written on a separate sheet or card. The ideas generated are sorted into various categories. This technique preserves the anonymity of the individual and overcomes many of the problems associated with brainstorming.

3.   INVENTION

Invention is defined as the creation of a new product, process or service. The legal definition of an inventor is someone who contributes to the *conception of the invention* of a complete and functional idea of how the invention will work (Weil and Snapper 1989, p. 61). Invention is more than envisioning the end result. The inventor provides direction and how to achieve the result. In order for an invention to be complete, the inventor must reduce the invention to practice. An invention not reduced to practice may not provide a strong patent (see Chapter Seven - *Technology Transfer*).

In order to protect an invention, it is important to reduce the invention to practice. The patentability of an invention has many implications. In many instances there is a lag from invention to commercial implementation. Table 4.1 shows the lag between invention and innovation, i.e., application.

However, as Figure 4.1 shows, the time between invention and innovation appears to be decreasing exponentially.

Table 4.1
Lag Between Invention and Innovation

Product or Process	Invention	Innovation
Crease-resisting fabric	1926	1932
Continuous steelcasting	1922	1952
Nylon, perlon	1927	1938
Jet engine	1928	1941
Penicillin	1928	1943
Sulzer loom	1928	1945
Synthetic light polarizer	1928	1932
Rocket	1929	1944
Cyclotron	1929	1937
Freon refrigerants	1930	1931
Polyethylene	1933	1937
Phototypesetting	1936	1954
Cinerama	1937	1953
Titanium	1937	1944
Xerography	1937	1950
Electronic digital computer	1939	1943
"Terylene" polyester fiber	1941	1955
Shell molding	1941	1948
Chordane, Aldrin, and Dieldrin	1944	1947
Long-playing record	1945	1948
Transistor	1947	1951
Oxygen steel making	1949	1952
Hovercraft	1955	1968
Semisynthetic penicillin	1957	1959
Wankle rotary piston engine	1957	1967
Moulton bicycle	1959	1963
Prevention of rhesus hemolytic disease	1961	1967

(Based on data contained in Sahal 1983)

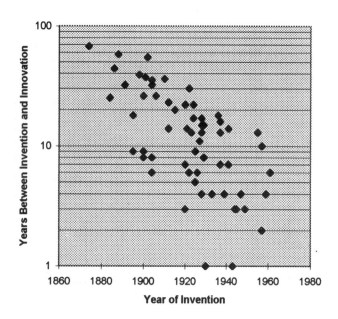

Fig. 4.1   Lag Between Invention and Innovation (1850 and 1970) (Based on data contained in Sahal 1983)

4.   INNOVATION

Innovation can be defined as the introduction of a new product, process or service into the market place (Edosomwan 1989, p. 3). Technological innovation can be defined as the process by which technological ideas are generated, developed, and transferred into new business products, processes and services that are used to make a profit and establish a marketplace advantage (Mogee 1993). All enterprises are faced with balancing and blending their organizational vector to efficiency and innovation (Clark and Staunton 1993, p. 127). Not being an innovating enterprise can have a high associated risk for the organization. Individual customers and markets expect periodic changes and improvements in products and services (Pearce and Robinson 1991, p. 234).

Figure 4.2 shows in general, according to Pearce and Robinson, that less than two percent of the innovative projects initially considered by the fifty-one companies they studied eventually reached the marketplace. Out of every fifty-eight new product ideas, only twelve passed an initial screening test that found them compatible with the enterprise's mission and long-term objectives. This study of Pearce and Robinson found only seven concepts remained after an evaluation of their potential, and only three survived development attempts. The remaining three innovations only produced two commercial products which had an apparent profit potential after test marketing and one was finally commercially successful. Some studies have even indicated higher failure rates.

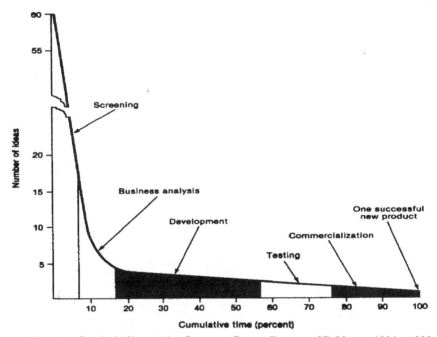

Fig. 4.2 Survival of Innovative Concepts (*Source*: Pearce and Robinson 1994, p. 233. Reprinted by permission.)

Clark and Staunton (Clark and Staunton 1993, p. 11-14) present a number of characterizations of innovations. *Generic innovations* are those innovations which create new technology-paradigms of clusters of innovations originating a new core process which cross-cuts many sectors and stages of production. Examples of generic innovations are the wheel, electricity and nuclear energy. *Epochal innovations* are subsets of generic innovations of importance whose introduction or rejection is confined to particular sectors of activity such as Polaroid film, Xerox copying, and Plexiglas. Innovations that introduce important alterations at the enterprise level are characterized as *altering innovations*. Examples of such altering innovations include groupware, the PC, and point-of-sale systems.

Entrenching innovations modify existing methods, but proceed along the same technological vector. These include the use of facsimile ("FAX"), introduction of the Federal Express system, and the ubiquitous voice mail systems. *Incremental innovations* do not provide new inputs, but the existing collections of inputs are reconfigured to achieve a higher output from the system. This type of innovation is exemplified by LAN and WAN systems, Management Information Systems ("MIS"), and Decision Support Systems ("DSS").

4.1 Sources of Innovation

Drucker (Drucker 1985, p. 35) has shown that innovation arises from unexpected occurrences, incongruities, process needs, industry and market change, demographic changes, changes in perception, and new knowledge. While these are potential rationales for the rise of an innovation, innovations can also come from different sources. Table 4.2 shows the functional sources of innovation. According to von Hipple (von Hipple 1988, p. 4), major product innovations in some fields are usually developed by the product users, i.e., the consumers of the innovation.

Table 4.2

Functional Sources of Innovation

Device	Innovation Developed By					
	User	Manufacturer	Supplier	Other	NA	Total
Scientific instruments	77%	23%	0%	0%	17	11
Semiconductor and printed circuit board process	67	21	0	12	6	49
Pultrusion process	90	10	0	0	0	10
Tractor shovel-related	6	94	0	0	0	16
Engineering plastics	10	90	0	0	0	5
Plastic additives	8	92	0	0	4	16
Industrial gas-using	42	17	33	8	0	12
Thermoplastics-using	43	14	36	7	0	14
Wire termination equipment	11	33	56	0	2	20

NA = data item was not available.

(Based on data contained in von Hipple, 1988, p.4)

There are many factors that influence the functional sources of innovation. According to von Hipple, the economic rents or temporary profits expected by potential innovators can be an excellent predictor of functional sources of innovation (von Hipple 1988, p. 5). This study by von Hipple leads to the conclusion that innovating enterprises could reasonably anticipate higher profits than non-innovating organizations. However, while in general terms this may be true, other research on pioneering in technology leads to the conclusion that the second entry enterprises are the organizations which achieve higher levels of profitability. An example of this is the major success of Microsoft® Corporation, an enterprise not considered an innovator, but having the user innovations developed outside the core enterprise.

4.2 Innovation Process

Figure 4.3 shows an innovation process or step model based upon Edosomwan's (Edosomwan 1989, p. 7) construct. The concept begins always with a new idea, either from an individual or group, that is influenced by some event in either the external or internal enterprise environment. Edosomwan's steps in the innovation process include: logical organization idea, refinement of idea, removing barriers, search for feasible solution, revising the idea, and implementation. This approach is basically another restatement of the scientific method or systems approach.

The innovation process requires support on various levels, national, enterprise and individual. According to Edosomwan, there are three levels which support the total innovation process (Edosomwan 1989, p. 6-10). Edosomwan considers *Level One* the Policy Formation or National and Enterprise Level. Level one has a number of different components which interact on a international, national and regional basis. Among these macro activities, according to Edosomwan, are the educational process in science and engineering, and the R&D funding and policies. These components are associated with the knowledge formation process. This level also includes the mechanism for local or regional technology delivery. External trade balance, incentivization, and the encouragement of joint ventures are also included in Level One since they form a basis for technology transfer. The incentives for the delivery and transfer of technology comprise the final components of Level One.

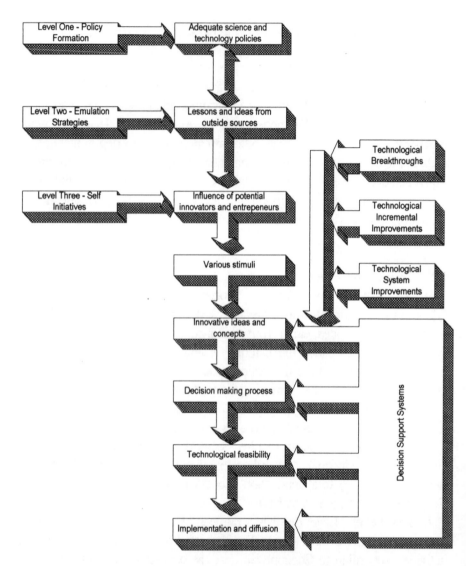

Fig. 4.3   Innovation Process Model (Adopted from Edosomwan 1989, p. 7)

Level Two, the Emulation Strategies or Enterprise Level, consists of identification of all distinct technologies, sub-technologies, and potential relevant technologies in the value chain. Assessment of the enterprise's capabilities in important technologies are part of the Enterprise Level. Cost making improvements, the selection of enterprise technological strategy (see Chapter Two - *Technological Strategy*), are included in this level.

The last level, i.e., *Level Three*, is the Self-Initiative and Individual Level which consists of those activities and structures which facilitate the innovation process at the basic technology level. Included in this are adequate facilities and infrastructure, the systems and structures that promote individual and organizational effectiveness and efficiency, and individual capability development.

The basic step or ladder model of innovation used by Edosomwan, illustrated by Figure 4.3, is now giving way to a *cyclical* type of innovation process. This innovation model emphasizes the importance of interactions among R&D, manufacturing, marketing and other corporate functions during the innovation process. This views innovation as totally an enterprise function versus Edosomwan's more global approach. Innovation can be viewed using both approaches. However, enterprises can rarely influence national or international factors, leading to innovation. Thus Edosomwan's model would be considered a national policy model while Mogee's model (Mogee 1993) is an enterprise level model.

This type of innovation process focuses on achieving continual, speedy improvements to existing technologies (Mogee 1993). The cyclic nature arises in the sense that it is driven by the product improvement cycle (see Chapter Five - *Technological Life Cycles and Decision Making*). This cycle often begins with customer needs. Also an enterprise may be working on multiple generations of a new product simultaneously. An example of this parallel innovation process is Intel's simultaneous development of the Pentium™ and P6 microprocessors.

4.3 Potential Cyclic Nature of Technological Innovation

The process of technological innovation involves complex relationships among a set of key variables:

- Inventions
- Innovations
- Diffusion paths
- Investment activity

The complex relations between these variables form a non-linear system with its underlying ramifications which can lead to unexpected and possible chaotic results. There is some evidence that technological innovations follow a cyclic process[3]. According to Rosenberg (Rosenberg 1994, p. 78-82), the first requirement for a technological theory of long cycles is a clear specification of causality among the various factors. Technological innovation may be at the center of both cyclical instability and economic growth, with the vector of causality moving clearly from the fluctuations in innovation to fluctuations in investment and from those to cycles in economic growth.

It is possible that a *strange technological attractor*[4] based upon the competitive forces derived from the Lotka-Volterra model could show this relationship in the technological phase space. Çambel's predator-prey model based on the Lotka-Volterra model may offer a means for understanding cyclical movements (Çambel 1993, p. 98-105).

Rosenberg implies that technological innovation leads to cycles of forty-five to sixty years in length, with long periods of expansion giving way to similarly extended periods of stagnation (Rosenberg 1994, p. 79). However, the various information and communication technologies are changing the rate of diffusion and thus the economic interactions. There is no doubt that the non-linearity of the technological innovation and diffusion process may under certain conditions lead to chaotic instability with the commensurate chaotic economic repercussions.

5.   MANAGEMENT OF INNOVATION

Technological innovation requires strong integration and orchestration of cross functional activities (Thamhain 1992, p. 436). The National Research Council ("NCR") defines the management of innovation as the *"linking of engineering, science and management disciplines to plan, develop and implement technological capabilities to shape and accomplish the strategic and operational objectives of an organization."* (Mogee 1993) The NRC listed, according to Mogee, eight primary needs in management innovation:

---

[3] Sometimes referred to Kondratiev Long Waves (Rosenberg 1994, p. 62).

[4] Strange technological attractor is based upon the concept that when a technological system in phase space is near an equilibrium point, it tends to evolve towards the state represented by that attractor or equilibrium point. A technological system can start from different sets of points in phase space, i.e., different technological developmental states and still end at the same attractor region. When the various paths the technological system can take to reach this attractor are represented by *fractals,* then the attractor is called a *strange attractor.*

- How to integrate technology into the overall strategic objectives of the enterprise (see Chapter Two - *Technological Strategy*).
- How to get into and out of technologies faster and more efficiently (see Chapter Six - *Enterprise Structure and Design*).
- How to assess or evaluate technology more efficiently (see Chapter Seven - Technology Transfer).
- How to accomplish technology transfer (see Chapter Seven - Technology Transfer).
- How to reduce new product development time (see Chapter Five - *Technological Life Cycles and Decision Making*).
- How to manage large, complex and interdisciplinary or inter-organizational projects and systems (see Chapter Six - *Enterprise Structure and Design*).
- How to manage the organization's internal use of technology (see Chapters Five and Six).
- How to leverage[5] the effectiveness of the technical professionals (see Chapter Six - *Enterprise Structure and Design*).

According to Mogee (Mogee 1993), innovation management education must be aimed at upper-level managers and managers of non-technical functions (e.g., finance and marketing) as well as managers of technological functions. Managers need a common ground including a series of skills for understanding of the process of technological innovation, and the competitive opportunities and risks it offers the enterprise. These skills include an awareness of the pervasive and complex nature of technological change (see Chapter 1 - *Technological Advancement and Competitive Advantage*). The technology manager must have a personal sensitivity to patterns of technology, market, and industrial development, and an appreciation for how technological change creates business opportunities (see Chapter Three - *Technological Forecasting*). Another tool the manager of technology must possess is an imagination to envision how available or potential technology can yield new products for future markets (see this Chapter), an appreciation of the need to protect the results of R&D, and innovation through strong intellectual property rights (see Chapter Seven - *Technology Transfer*).

---

[5] Leverage means that a relatively small change in the effectiveness of technical professionals will result in a large change in the effectiveness of the innovation process within an enterprise.

Successful technological innovation often consists of building a unified multifunctional team committed to the innovative implementation of a technological strategic plan (see Chapter Two - *Technological Strategy*). The management structures, rationales, actions and intra-managerial politics affect the technological innovation performance of an enterprise (Webb 1993). In some studies, enterprise innovations were undermined by fear of failure at top management levels. This results in a pattern of defensive avoidance and actions focusing on the control of costs, as a tangible variable to the exclusion of all other considerations. The inability of management to reflect on and adapt their action can possibly be attributed to the general ethos of aggressive *macho* management in innovation (Webb 1993).

5.1 Innovative Organizations

Mintzberg's concept of innovative organizations, also referred to as *adhocracy*, holds for technological and non-technological enterprises (Mintzberg and Quinn 1991, p. 732). Innovation requires a structure that is capable of coalescing individual technological and non-technological performers into an effectively functioning project team. According to Mintzberg, the innovative organisation cannot rely on any form of standardized co-ordination. The enterprise must avoid bureaucratic structure, including structures that result in sharp division of labor, extensive unit differentiation, highly formalized behavior, and an emphasis on planning and control systems with a high degree of flexibility.

In the current restructuring of many organizations, managers of technological innovation need to manage in the light of enterprise downsizing, restructuring and transient senior management (Day et al. 1994). Studies indicate that a new management format will be needed for twenty-first century enterprises to succeed. Accordingly, enterprises will not be able to remain globally competitive and profitable enough without finding a success formula for better managing innovation.

5.2 Modes of Enterprise Innovation

Studies indicate that there are several modes of enterprise innovation (Miller and Blais 1993). In the *science-based product innovation* modes the enterprises are oriented toward the internal development of new products based on substantial R&D investment (five to twelve percent of sales). These enterprises are technological leaders and pacesetters. The *entrepreneurial fast track experimentation innovation* enterprise mode is characterized by a high degree of experimentation to develop a continuous flow of improved products and production. This mode is also characterized by the rapidity of development of new products and services, which often have high associated risks. Enterprises in the *global cost leadership innovation* mode are process innovators, with strength derived from the development or acquisition of new processes. These are high quality - low cost driven enterprises.

Reliance on information technology and process adoption is another mode of enterprise innovation. These enterprises are innovators who incorporate new ideas stemming from information technology or other engineering enterprises. The banking service industry is an example of this type of innovative enterprise mode, where internal staff may lack a sufficient high degree of technical knowledge, but rely on the hardware manufacturers to provide the technology upon which their new services will be based and which must be relied upon to work effectively.

6.   THE ACQUISITION PROCESS

The verb, *acquire*, is defined by the *Random House Dictionary of the English Language* (Stein 1966) as:

> *"1. to come into possession of; get as one's own;*
> *2. to gain for oneself through one's actions or*
> *efforts."*

The acquisition process is the process of acquiring a new product, process or service by efforts of individuals or an enterprise. This process can be conducted either internally or externally to the enterprise. The particular acquisition strategy chosen depends on a number of different factors. The technological strategy of the enterprise will determine the factors for choosing the specific technology acquisition alternative (see Chapter Two - *Technological Strategy*). The decision also depends on market or operational factors.

6.1   First Movers or Market Pioneering

An enterprise which first produces a new product, uses a new process, enters a new market, or provides a new service may accrue a long-term competitive advantage (Kerin et al. 1993). Kerin et al. indicate that these *first movers* have higher market shares than early followers, who in turn have higher market shares than later entrants. However, just getting there first is not the entire answer, since it is not sufficient to achieve cost and product advantages over rivals that result in maximizing market share and accruing large economic rents[6]. To maximize the advantage of *first movers* or *market pioneering*, according to Kerin et al., enterprises must possess certain competencies and capabilities, which include technological foresight, perceptive market research, skillful product and process development capabilities, and marketing acumen.

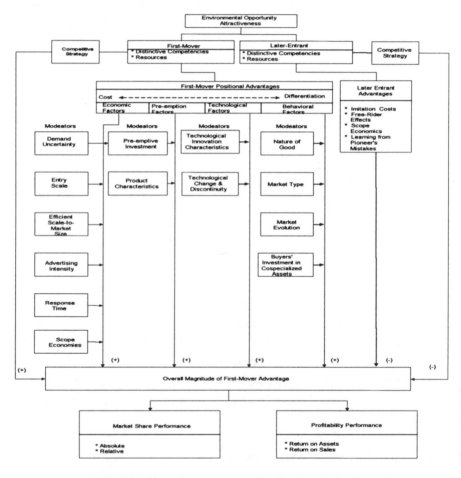

Fig. 4.4 First Mover Advantage Model (*Source*: Kerin et al. 1993, © 1993 IEEE. Reprinted by permission.)

---

[6] Profits that more than cover the opportunity cost of capital are known as *economic rents*.

Some firms are successful leaders or *first movers* while others do better as followers. This *market pioneering* only provides opportunities: it is not the only factor in achieving market dominance. Unless an enterprise has expertise, resources, and creativity to exploit these opportunities, it will be unable to achieve dominance and produce sustainable competitive advantage. In some instances, organizations have the creativity, but lack the expertise or resources to achieve the competitive advantage. Other firms might have the internal resources, but lack the expertise needed to become the *market pioneer*. These enterprises might decide to acquire a successful early entrant which does not have sustainable resources to capitalize on their *first mover* advantage. Figure 4.4 presents a model of the first mover advantage.

7. INTERNAL ACQUISITION

7.1 New Products, Processes and Services

The success of new products, processes or services depends upon a number of factors. The guidelines for acquiring a successful new product, process or service have been researched by Cooper and Kleinschmidt (Cooper and Kleinschmidt 1993). An enterprise hoping to succeed with a new technological market entrant must seek to differentiate its new product, process or service. This differentiation, in the eyes of the customer, must be advantage driven. Success depends upon management and leadership team acceptance of a sharp and early definition of the technological product, process or service. This early definition includes target market, concept, positioning, benefits, and features and specifications. Enterprise to succeed should also organize around a cross-functional, also known as multifunctional, new team. These multifunctional teams must have a strong empowered team leader, accountability, and top management support. Like a military maneuver, the technological enterprise should attack the market from a position of strength. Synergy is the key, and the enterprise must make itself aware of the stochastic events which can rapidly change the technological landscape. Familiarity with various aspects of the developmental process and market is another important element. The focus enterprise is on *quality of execution*, in every step of the definition and development process.

According to Cooper and Kleinschmidt (Cooper and Kleinschmidt 1993), the new product, process or service internal acquisition process must be preoccupied with seeking and achieving real advantages from a customer perspective. Forcing customer interfacing throughout the entire definition and development process is part of the successful acquisition process. It is also important that the enterprise rely on the *voice of the customer* as vital input into the design and not just as after-the-fact confirmation. The process should build in a *user-needs-and-wants* study via face-to-face interviews with potential users. An important element for the manager of technology is to have the technical staff in direct and frequent contact with the customer. This customer focus is vital to insure that the resulting technological product, process or service meets the needs of the market. Included is the continual customer feedback during development, testing and validation of the product, process or service throughout the development stage.

The technological acquisition process incorporates a thorough competitor and competitive analysis to determine what should be built into the new item and how competition should be managed. Also, successful technological enterprises build in a concept test to check and confirm the proposed new concept or design and to gage interest, liking preference and purchase intent. This would include the incorporation of the traditional tests which occur after the development phase, such as use of field trials in order to verify functions and establish purchase intent, pre-test market, test market, and trial sell.

Building in a definition point before development begins where the management and project team buy into the product, process or service definition is important. A mandated multifunctional team approach has been shown to be an important element of successful enterprise technological acquisition processes. This would require a strong, empowered project team leader with formal authority. The leadership team of the enterprise should enforce team accountability for the project, where the team leader and team are fully accountable for the entire process. The technological team leader, in turn, insures top management support for the project; this should be built into the early stages of the decision and selection process. When the technological process cuts across functional boundaries, the manager of technology forces a *true* multifunctional process. The leadership team of the enterprise provides for sharper project selection at the decision points by using selection criteria that are critical to success, including advantages, synergy, familiarity, market attractiveness, and the competitive situation. Above all, the acquisition process must have a preoccupation with quality of execution at every step of the process, i.e., build in quality check points.

7.2 Screening New Products, Processes or Services

Enterprises are increasingly dependent on new products, processes or services. Over forty percent of sales of U.S. firms come from new entrants (Cooper and Kleinschmidt 1994). However, only one development project in four or five becomes a commercial success (see Figure 4.2). Half the resources devoted by U.S. industry to innovation and development result in unsuccessful products, processes or services. Most new development projects are characterized by serious deficiencies such as errors of omission, poor quality of execution, questionable project selection, and other deficiencies. According to Cooper and Kleinschmidt (Cooper and Kleinschmidt 1994), most enterprise leadership (sixty-three percent) is *somewhat* or *very disappointed* in the results of their enterprise's new entrants' efforts.

Table 4.3 shows the impact of new entrant success factors. Figure 4.5 shows a new product (process or service) process.

Table 4.3
Impact of New Entrant Success Factors

Success Factor	Bottom 20 %[1]	Middle 60%	Top 20%	Difference: Top 20% vs. Bottom 20%
Product Advantage	18.4%[2] (11.6%)[3]	58.0% (32.4%)	98.0% (53.5%)	79.0% (41.9%)
Quality of Pre-Develop. Activities	31.3% (20.8%)	68.1% (35.6%)	75.0% (45.7%)	43.7% (24.9%)
Early Product & Project Definition	26.2% (22.9%)	64.2% (36.5%)	85.4% (37.3%)	59.2% (14.4%)
Quality of Marketing Activities	32.5% (24.6%)	66.1% (34.6%)	71.1% (42.1%)	38.6% (17.5%)
Quality of Technological Activities	29.7% (21.7%)	64.8% (34.6%)	75.6% (42.6%)	45.9% (20.9%)

Notes:
1. Projects split into 3 groups on each factor; e.g., for Product Advantage, those 20% of projects with high product advantage vs. Lowest 20%.
2. Percent successful (% exceeding financial criteria for success).
3. Market share, end of year 2.

(*Source*: Cooper and Kleinschmidt 1994, © 1994 IEEE. Reprinted by permission.)

134

The formal product, process or services development system of Cooper and Kleinschmidt (Cooper and Kleinschmidt 1994) incorporates the success factors of improved teamwork, less recycling and rework, improved success rate, better launch, earlier detection of failures, and a shorter development cycle. The development of a successful technological market entrant should incorporate these factors.

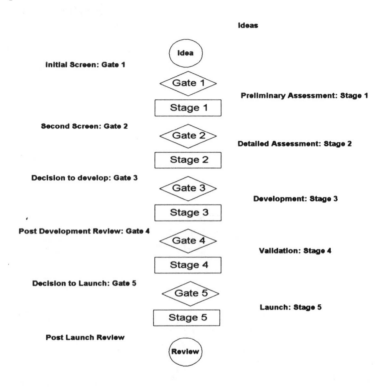

Fig. 4.5 New Product, Process or Service Model (*Source*: Cooper and Kleinschmidt 1994, © 1994 IEEE. Reprinted by permission.)

7.3 Development Process

The organizational environment of the future is becoming highly stochastic and unpredictable. Enterprises must adapt to changing conditions and be more responsive to specific problems to be solved (Thamhain 1992, p. 276-277). Success in the future environment will require enterprises to:

- consist of temporary systems
- organize around problems
- use multifunctional teams

New product, process or service development can be enhanced by a horizontal enterprise structure (Byrne 1993). These horizontal enterprises should be organized around processes, not tasks, and have a flat hierarchical structure. Successful horizontal enterprises use teams to manage everything. i.e., a *team village* structure within the enterprise. Successful enterprises should be customer driven, employ maximization of supplier and customer contact, and have informed and trained employees. This requires the replacement of rigid management systems with more flexible temporary management systems (Byrne 1993). However, these new structures will continue to operate within a functional framework in order to provide the required stability and efficiency.

As the competitive environment has changed, many technological enterprises have chosen to focus resources on refining and enhancing their traditional strengths in R&D and have neglected other processes required to bring a product, process or service to market (Pisano and Wheelwright 1995). Enterprises such as Intel, Hewlett-Packard, and Texas Instruments have built, during the 1980s and 1990s, a unique and sustainable competitive position by creating capabilities (processes) that have resulted in faster, more frequent, more productive and more effective new market entries. According to Pisano and Wheelwright (Pisano and Wheelwright 1995), what goes on before the introduction of a technological product, process or service is more important than what goes on during and after the introduction.

7.4 Multifunctional Teams

The internal acquisition of technological products, processes or services involves techniques to effectively lead and integrate multidisciplinary activities (Thamhain 1992, p. 279). A high degree of integration and co-operation are required between the functions of R&D, marketing and manufacturing to achieve effective new development performance (Wilemon 1995). It is important that enterprises construct an internal environment where multifunctional teams can successfully operate. True multifunctional integration occurs at the working level where communication is very critical.

A multifunctional team, according to Wilemon (Wilemon 1995), also known as cross-functional team ("CFT"), is defined as:

> *"Members of different departments and disciplines brought together under one manager*

to make development decisions and enlist
support for them throughout the organization."

The development of multifunctional teams evolved because traditional approaches for developing new technologically based products, processes or services were not effective. The cause of this ineffectiveness was limited internal integration, i.e., among R&D, manufacturing and marketing, and limited external integration, i.e., customers, suppliers, partners and technologies. Other contributing factors were poorly executed or ignored developmental steps, combined with the important initial steps being rushed. These factors produced technical problems later in the developmental process. Low information sharing among enterprise sectors and limited learning across projects and organizations were others.

The benefits of multifunctional teams include:

- Synergy, i.e., more than the sum of the parts.
- Improved customer focus and satisfaction.
- Reduced lead time to market for new products, processes or services.
- Higher quality decisions and work.
- Fewer communication breakdowns.
- Enhanced organizational learning.

However, according to Wilemon (Wilemon 1995), multifunctional teams have limitations because they can be slow and painful for an enterprise. Team collaborations are often tentative, fragile, threatened by confusion, stressful, conflicting and skeptical, and team member contributions are quite variable. There is also the paradox of preserving differences among team members while attempting to integrate the differences into a whole. And there is definite team tension as members adjust to becoming team members.

According to Wilemon (Wilemon 1995) and Thamhain (Thamhain 1992, p. 279), multifunctional teams can be enhanced through a number of actions. By breaking large projects into smaller ones, teams of more than eight members have been found to have significant interface problems. Figure 4.6 presents the building blocks for an effective multifunctional team. Limitation of team size to eight or fewer members will reduce interface problems. It is important to take a proactive stance toward interface problems by openly discussing them and seeking resolutions and eliminating small problems early. Involving all parties early in the project life cycle also enhances team performance. Promoting and maintaining dyadic relationships, based upon complementary skills, become the kernel of larger circles of responsibility.

By making open communication the responsibility of every team member, performance of the multifunctional team is enhanced. Interlocking task forces by creating a multifunctional top-level steering committee, and clarifying the decision-making authorities can assist performance. Altering the basis for allocating responsibility for goal achievements from controllers to direct participants, e.g., change managers from resource controllers to resource suppliers, is another technique. An important factor is empowering teams and not individual team members. The team is like a family when the team is empowered rather than the individual team members. It is important to give professionals the ability to choose their assignments within the structure. Successful multifunctional teams have a balanced leadership with membership, i.e., they do not make it a hierarchical structure , they use a *flat structure*. All team members should be trained for team work. According to Wilemon (Wilemon 1995), the use of multifunctional teams has resulted in: more products, processes and services; faster realization times; higher product quality; and the creation of products, processes and services which better reflect the needs of both the customers and the enterprise.

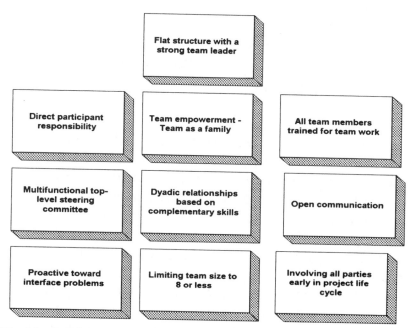

Fig. 4.6   Building Blocks to Successful Multifunctional Teams

8.   EXTERNAL ACQUISITION

An enterprise might externally acquire or develop its technological products, processes or services due to lack of sufficient internal resources, i.e., human, financial or physical. Another factor leading to external acquisition is the lack of core competencies to deal with complex technological developments, and the lack of risk taking ability, i.e., if another organization fails, little or no blame will accrue to the enterprise or management.

From a marketing perspective, external acquisition may be undertaken to provide a stealth market development. This type of acquisition also provides a means of acquiring specific skills that do not exist within the technological enterprise. As many technology enterprises have demonstrated, external acquisition provide a means to develop a strategic relationship.   Another reason for external acquisition is when a technological activity is not within the specific organizational charter; this is the preferred method of the Government of the United States, which utilizes the private sector to obtain technological developments, including government owned contractor operated ("GOCO") laboratories.

This process for an enterprise in acquiring technology externally can be accomplished by:

- Outsourcing
- Strategic Alliances
- Collaborative Research and Development
- Enterprise Acquisition

8.1   Outsourcing

The drive for enterprises to become more competitive, by reducing cost, has led to outsourcing of technological developments. A financial comparison of outsourcing versus further capital expenditures usually highlights the risking of further expenditures on operations with unsatisfactory results (Bettis et al. 1993). Far Eastern technology enterprises in Korea, China, Taiwan, India and Singapore have been the major beneficiaries of technological outsourcing from western organizations. These enterprises usually provide exemplary services at lower cost. The option of outsourcing becomes easier and easier as the outsourcing enterprise's expectations on delivery and quality are often exceeded. Thus an outsourcing technological attractor is formed which could lead to manufacturing decline in nations which enter this cycle, as illustrated by the model shown in Figure 4.7. It is important for both nations and individual enterprises to be aware of this potential cycle and take the necessary steps to improve their long term competitive positions.

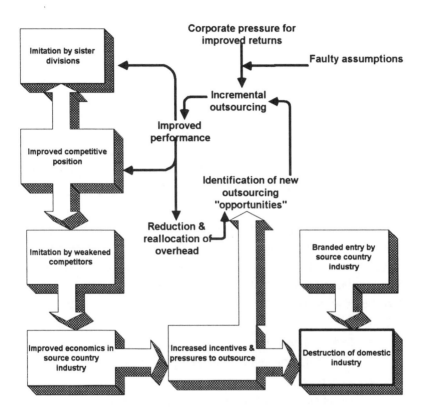

Fig. 4.7 Outsourcing Technological Attractor Model (*Source*: Bettis et al. 1993, © 1993 IEEE. Reprinted by permission.)

Outsourcing can prove to be a valuable developmental tool if viewed in a strategic fashion. Treating outsourcing decisions, strategically, implies an in-depth understanding of the core competencies in which the enterprise intends to build its future competitive advantage. Therefore, outsourcing should focus on technological areas different from the enterprise's core competencies (Bettis et al. 1993). Increasing outsourcing to include areas closer to core competencies increases strategic risk. The impact of outsourcing decisions on continued skill and competence development is another consideration. The intention of an enterprise to become a supplier is an important consideration. For example, use of suppliers whose intentions appear to potentially involve acquiring your enterprise technology, market knowledge and core competencies would be a high risk decision.

Any savings from outsourcing should be applied to increasing core competencies to sustain competitive advantage. Outsourcing should be used to learn from the supplier. These learning opportunities are symmetrical, i.e., both parties to the contract learn (see Chapter Six - *Enterprise Structure and Design*).

8.2   Strategic Alliances and Joint Ventures

According to Hamel and Prahalad (Hamel and Prahalad 1994, p. 167), the use of strategic alliances is one method of *borrowing* the resources of other enterprises.   Long-term strategic alliances allow enterprise insight into competencies that are deeply buried within the fabric of a partner.   In the majority of strategic alliances, the dominant pattern of activity is the exchange of technology for technology (Simon 1994).   According to Simon, strategic alliances usually emphasize a bilateral flow of technology, while joint ventures tend to emphasize a unilateral flow of technology.   Strategic alliances have caused the flow of technology before the technology's full market value has been exploited.   Technological strategic alliances can take many forms:

- technical exchange and cross-licensing;
- co-production and marketing agreements;
- joint product development programs;
- stand-alone joint venture enterprises with equity distribution among participants.

Strategic alliances are formed to combine the superior design technology of one of the partners with the efficient manufacturing of the other partner. This type of alliance results in the diversifying the risks inherent in developing new products, processes or services.

The strategic intent behind each partner in forming the alliance is very important.   According to Pucik (Pucik 1994), the change from competitive rivalry to collaboration is merely a tactical adjustment aimed at specific market conditions.

A potential competitive relationship between partners distinguishes strategic alliances that involve competitive collaboration from more traditional complementary ventures (Hamel and Prahalad 1994, p. 213). The following is a comparison between co-operative and competitive collaboration:

Co-operative Collaboration

- Feasibility
- Desirability
- Long-term win/win outcomes

Competitive Collaboration

- Strategic market intent
- No long-term win/win outcomes likely

Table 4.4 illustrates the barriers to organizational learning in strategic alliances. Kumaroswamy (Kumaroswamy 1995) presents a strategy for assessing strategic alliances and joint ventures.

Table 4.4

Barriers To Organizational Learning In Strategic Alliances

Functional Area	Principal Barriers
Strategy Formation	Short-term and static planning horizon No appreciation of incremental learning Strategic intent not communicated Low priority of learning activities Fragmentation of the learning process
Human Resource Allocation	Lack of involvement of the human resource function Insufficient lead-time for staffing decisions Resource-poor human resource strategy Surrendering control over the human resource function Staffing dependence on the partner
Management	Low quality of staff assigned to the alliance Lack of cross-cultural competence Unidirectional personnel transfer Career structure not conducive to learning Poor climate for transfer of knowledge
Control	Responsibility for learning not clear Short-term performance measures Limited incentives for learning Tolerance of learning barriers Rewards not tied to global strategy

(*Source*: Based on data contained in Pucik 1994)

8.3 Collaborative Research and Development

Collaborative research and development among universities, industry and government laboratories is increasing. Laws in the United States have encouraged strategic alliances and co-operative R&D across organizational boundaries[7], fostering the cross-organizational collaboration by the formation of research and development consortia (Gibson et al. 1994). Trust among the partners of collaborative research and development is an essential prerequisite for success (Haüsler et al. 1995). Industrial co-operation can be considered very important to an enterprise's survival over the long term.

Haüsler et al (Haüsler et al. 1995), proposed a *cascade model* to describe a process of collaborative research and development. In this model, initially, a large number of technologists from industry and academia meet in loosely coupled networks to discuss future technological developments in a specific area. At the end of these meetings, a smaller group of researchers from a small number of enterprises and universities co-operate in a government-industry sponsored research and development project to successfully prove the technology. This general co-operative research and development process can be summarized as:

- Stage 1: Establishment of a Scientific-Technology Network
- Stage 2: Agreement of R&D Collaboration
- Stage 3: Collaboration in R&D Project

8.4 Enterprise Acquisition

An approach for developing a new technology is to acquire another enterprise which may have essential technology elements. These enterprises may have developed a critical technological element such as devices, processes, products, or services which may be a key element in the technological strategic process. The target enterprise may have existing core competencies which will be needed in the development process, or strategic assets such as specific manufacturing facilities. The technology enterprise could also have a dominant market position in which the enterprise wants to position the technology, such as in the 1995 acquisition of Lotus Corporation by IBM. Acquisition of enterprises has always been an important activity in the United

---

[7] Stevenson-Wydler Innovation Act of 1980, Bayh-Dole Act of 1980, National Co-operative Research Act of 1984, and Federal Technology Transfer Act of 1986.

States. The dominant motives for acquisition can be divided into (Chakrabarti et al. 1995):

- Market and marketing-related motives
- Cost reduction
- Technological know-how
- Financial
- Political

The success of a technological acquisition, according to Chakrabarti et al. (Chakrabarti et al. 1995) can be measured in terms of technical success and economic success. Technical success is negatively influenced by the technological uncertainty of the scientific and technology field in which the partners of the acquisition are involved. Matching the size of the potential partners is very important to the technical success of the acquisition process. Problems seem to arise when small enterprises are acquired by larger enterprises due to cultural differences.

Acquisition of an enterprise in crisis may seem to be a method of acquiring technology inexpensively, but research indicates it may lead to acquisition of the crisis. The timeliness of the integration process is very important, i.e., the degree of fulfilling the expectations on a time scale. Another important factor is the open, confident and permanent interaction and communication of the R&D teams.

Economic success is closely related to the enterprise acquisition process. Chakrabarti et al. (Chakrabarti et al. 1995) found that experience with prior enterprise acquisition has a negative impact on economic success. Matching the size of the enterprise has a positive influence on technological success and an indirect negative influence on economic success. Large firms acquiring small firms have a negative technological impact, but a positive economic impact. According to Chakrabarti et al., economic success was not strongly connected with technical success in the acquisition process. Conflicting technological philosophies between the partners of the acquisition have the strongest negative impact on economic success, but have little effect on technological success. Technological conflicts have a delayed impact on economic success due to the interval between development and marketing. Profit appears later and under other circumstances than technical success

If the co-operation reveals differences in the technological philosophy, the success will decrease. Negative effects found by Chakrabarti et al. (Chakrabarti et al. 1995) can be overcome by deliberate development of clear

144

goals. The use of internal and external experts, and to a certain extent careful decision making, will also reduce negative effects. The autonomy of the acquired enterprise is also a favorable condition for economic success, and it is important that there be open, confident and mutual communication. It is appropriate management which makes a enterprise acquisition successful.

REFERENCES

Bettis, R. A., Bradley, S. P., and Hamel, G. (1993). "Outsourcing and Industrial Decline." *IEEE Engineering Management Review*, 21(Winter), 85 - 93.

Byrne, J. (1993). "The Horizontal Organization." Business Week, 76 - 81.

Çambel, A. B. (1993). *Applied Chaos Theory: A Paradigm for Complexity*, Academic Press, Inc., Boston, MA.

Chakrabarti, A., Hauschildt, J., and Süverkrüp, C. (1995). "Does It Pay to Acquire Technological Firms." *IEEE Engineering Management Review*, 21(Spring), 35 - 41.

Clark, P., and Staunton, N. (1993). *Innovation in Technology and Organisation*, Routledge, London, UK.

Cooper, R. G., and Kleinschmidt, E. J. (1993). "Uncovering the Keys to New Product Success." *IEEE Engineering Management Review*, 21(Winter), 5 -18.

Cooper, R. G., and Kleinschmidt, E. J. (1994). "Screening New Products for Potential Winners." *IEEE Engineering Management Review*, 22(Winter), 24 - 30.

Day, G. S., Gold, B., and Kuczmarski, T. D. (1994). "Significant Issues for Future of Product Innovation." *IEEE Engineering Management Review*, 22(Winter), 2 - 7.

Drucker, P. (1985). *Innovation and Entrepreneurship*, Harper and Row Publishers, New York, NY.

Edosomwan, J. A. (1989). *Integrating Innovation and Technology Management*, John Wiley & Sons, New York, NY.

Gibson, D. V., Kehoe, A., and Lee, S.-Y. K. (1994). "Collaborative Research as a Function of Proximity, Industry and Company: A Case Study of an R&D Consortium." *IEEE Transactions in Engineering Management*, 41(August), 255 - 263.

Hamel, G., and Prahalad, C. K. (1994). *Competing for the Future*, Harvard Business School Press, Boston, MA.

Häusler, J., Hohn, H. W., and Lütz, S. (1995). "Contingencies of Innovative Networks: A Case Study of Successful Interfirm R&D Collaboration." *IEEE Engineering Management Review*, 23(Spring), 42 - 55.

Howard Jr., W. G., and Guile, B. R. (1992). *Profiting from Innovation: The Report of the Three-Year Study from the National Academy of Engineering*, The Free Press, New York, NY.

Kerin, R. A., Varadarajan, P. R., and Peterson, R. A. (1993). "First-Mover Advantage - A Synthesis, Conceptual Framework and Research Proposition." *IEEE Engineering Management Review*, 21(Winter), 19 - 33.

Kumaroswamy, M. M. "Technology Exchange as a Corner-stone of Cross-Cultural Engineering Management." *IEEE Annual International Engineering Management Conference*, Singapore, 315 - 320.

Miller, R., and Blais, R. A. (1993). "Modes of Innovation in Six Industrial Sectors." *IEEE Transactions in Engineering Management*, 40(August), 264 - 273.

Mintzberg, H., and Quinn, J. B. (1991). *The Strategy Process: Concepts, Contexts, Cases*, Prentice-Hall, Inc., Englewood Cliffs, NJ.

Mogee, M. E. (1993). "Educating Innovation Managers: Strategic Issues for Business and Higher Education." *IEEE Transactions in Engineering Management*, 40(November), 410 - 417.

Pearce, J. A., II , and Robinson , R. B., Jr. (1994). *Strategic Management: Formulation. Implementation, and Control*, Richard D. Irwin, Inc., Homewood, IL.

Pisano, G. P., and Wheelwright, S. C. (1995). "The New Logic of High-Tech R&D." *Harvard Business Review*(September-October), 93 -105.

Porter, A. L., Roper, A. T., Mason, T. W., Rossini, F. A., and Banks, J. (1991). *Forecasting and Management of Technology*, John Wiley & Sons, New York, NY.

Pucik, V. (1994). "Technology Transfer in Strategic Alliances: Competitive Collaboration and Organizational Learning." Technology Transfer in International Business, T. Agmon and M. A. von Glinow, eds., Oxford University Press, New York, NY, 121-138.

Rosenberg, N. (1994). *Exploring the Black Box*, Cambridge University Press, New York, NY.

Sahal, D. (1983). "Invention, Innovation, and Economic Evolution." *Technological Forecasting and Social Change*, 23, 213-235.

Simon, D. (1994). "International Business and the Transborder Movement of Technology: A Dialectic Perspective." Technology Transfer in International Business, T. Agmon and M. A. von Glinow, eds., Oxford University Press, New York, NY, 5-28.

Stein, J. (1966). *The Random House Dictionary of the English Language*, Random House, New York, NY.

Thamhain, H. J. (1992). *Engineering Management: Managing Effectively in Technology-Based Organizations*, John Wiley & Sons, New York, NY.

von Hipple, E. (1988). *The Sources of Innovation*, Oxford University Press, New York, NY.

Webb, J. (1993). "The Mismanagement of Innovation." *IEEE Engineering Management Review*, 21(Summer), 11 - 20.

Weil, V., and Snapper, J. W. (1989). *Owning Scientific and Technical Information*, Rutgers University Press, New Brunswick, NJ.

Wilemon, D. "Cross-Functional Teamwork in Technology-Based Organizations." *IEEE Annual International Engineering Management*, Singapore, 74 - 79.

DISCUSSION QUESTIONS

1. Comment upon the senior management role in innovation in the following organizations:
    - Governmental agency - Enterprise Level
    - Governmental laboratory - R&D Organizational Level
    - Large multinational enterprise - Enterprise Level
    - Medium size technology company - Enterprise Level

- New entrepreneurial technology enterprise - Enterprise Level
- Corporate laboratory - R&D Organizational Level

2. What were the innovative characteristics behind each of the following developments:
   - Aeroplane
   - Transistor
   - Computer
   - Personal Computer

3. Describe an ideal innovative enterprise and identify its inherent characteristics.

4. How does the enterprise you work for measure up to this ideal organization ?

5. How should technical professionals be managed in a high-technology environment to stimulate innovation?

6. How can an innovative technology manager improve the confidence and trust in the organization ?

7. Recently IBM acquired Lotus, comment upon this acquisition in reference to the material discussed in this chapter.

8. Write about the differences between the private sector and the Governmental sector in the acquisition of technology.

9. Outsourcing has become an important part of the technological developmental process; discuss the pros and cons of this strategy in reference to a small enterprise, medium enterprise and large multinational enterprise.

10. Strategic alliances are important elements in developing technology; discuss this concept from the point of view of a small, medium and large enterprise.

CHAPTER 5

Technological Life Cycles and Decision Making

1. INTRODUCTION

All technologies progress through a series of life stages or cycles. The acquisition process is the first stage of the total life of any technology. Management of technology requires many decisions to be made throughout the total life cycle of producing a new technological product, process or service. However, due to the stochastic nature of the internal and external environment in which the process is embedded, the manager is faced with risk and its mitigation.

2. TECHNOLOGICAL LIFE CYCLES

Life cycles are a way of expressing the dynamic nature of various systems (Cleland and King 1983, p. 237). The technology which is proposed as the solution to a particular problem moves from state to state as it evolves from a creative solution, to final implementation, to phase out. A manager of technology is faced with various views of life cycles of technology. While all the various views of technological life cycles are considering the same development of the technology, each is from a different perspective. It is like looking at the same scene, but from windows on different floors of a building. Each view presents its perspective with emphasis on particular elements of the scene. Life cycle views which a manager of technology must consider include:

- Technology life cycle
- Product, process, or service life cycle
- System life cycle
- Project life cycle

- Organizational life cycle
- Business life cycle

Table 5.1 shows a summary of the various life cycle stages that a manager of technology must deal with during the total process of bringing a technology from its conceptual stage to eventual phase-out. Few technological developments are totally independent of other technologies or systems. In some instances, the technology that a manager is responsible for developing will be in the middle of the research and development portion of the technology life cycle while the product in which the technology is embedded is in the growth stage. An example of this often occurs in the personal computer arena, i.e., the technology may be replacing another technology, e.g., 32 MB[1] for 16 MB DRAM[2] chips. In complex systems, such as the development of a new aircraft like the Boeing 777, the overall system in which the product serves as one of the components may be in the utilization stage while the project is in the operational stage. Concurrently, the organization producing the product can be in a collaborative stage facing a series of complex problems and the overall business life cycle for the system in which the product   is used may be in the decline portion of the cycle.

The technology manager is faced with problems and opportunities due to this complexity. As in all technological complex systems, underlying non-linearity conditions exist which can cause a chaotic and potentially destabilization of all the cycles.

Martino (Martino 1993, p. 9) divides the technology life cycle into the following phases (see Chapter One - *Technological Advancement and Competitive Advantage*):

- Scientific findings
- Laboratory feasibility
- Operating prototype
- Commercial introduction or operational use
- Widespread adoption
- Diffusion to other areas
- Social and economic impact

---

[1] MB = megabyte,

[2] DRAM = dynamic random access memory

Table 5.1
Summary of Life Cycle Phases

General	Technology (Martino 1993, p. 9)	Product/Process/Service (Cleland and King 1983, p. 238-239)	System (Blanchard and Fabrycky 1990, p. 17-19)	Project (Cleland and King 1983, p. 246)	Organizational (Cleland and King 1983, p. 237-238)	Business (Howard Jr. and Guile 1992, p. 11-13)
Birth	Scientific Findings	Establishment	Acquisition Conceptualize	Preparation or Initiation	Creativity	Emergence
Early Growth	Laboratory Feasibility	Growth	Design and Development	Implementation	Directive	
Later Growth	Operating Prototype		Production/ Construction		Delegation	Growth
Stable Maturity	Commercial/ Operational Use	Maturation	Utilization Use	Operation	Coordinative	Maturity
Unstable Maturity	Diffusion to other areas				Collaborative	
Phase-out	Social and Economic Impact	Declining Use	Phase-out		Dissolution/ Acquired/ Merged	Possible Decline
			Disposal			

The first three phases of Martino's technology innovation life cycle correspond to the research and development phase of a technological development. As a technology enters the commercialization or operational use stage, it also enters other stages which are associated with the life cycle of the product, process, or service in which the technology is embedded.

2.1 Product, Process or Service Life Cycle

A product, process or service in which a technology is embedded moves through a life cycle in sales, market penetration or operational use. Figure 5.1 shows a product life cycle in terms of sales revenue. According to Cleland and King (Cleland and King 1983) the product life cycle starts with the establishment phase where the product is introduced into its intended market. The establishment phase is the most critical because if the product does not achieve sufficient market penetration, it will likely be withdrawn from the market. The computer software market is an example where programs are introduced and then rapidly fade from the market. If a product, process, or service has initial success, growth in market penetration may follow.

During the growth phase of the life cycle, the technology may begin to make significant market penetration and achieve a following, thus insuring its survival into the next phase. An example of this is the rapid rise of Netscape Corporation's World Wide Web ("WWW") Internet browser in the 1995 - 1996 time period. After being introduced in late 1994, this software achieved over eighty percent market penetration by late 1995. At some point, the product, process or service achieves a position were only minimal additional sales will be made. This plateau in market penetration may be sustained for a considerable interval. However, without additional performance improvements a decline in sales will occur.

The first three phases may be depicted by a form of S-shaped curve (see Chapter Three - *Technological Forecasting*). Most products, processes or services display these dynamic characteristics, although some may have a life cycle which is so long or short that some phases may not be readily distinguishable. With some products, process or service, the maturation phase is long and the decline portion of the life cycle is very gradual.

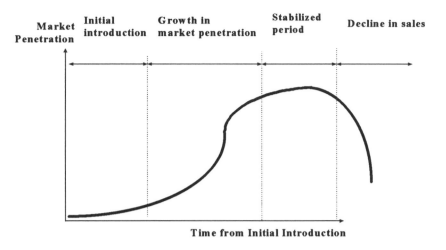

Fig. 5.1 Life Cycle for Product, Process or Service

The manager of technology will be faced with embedding the technology within a product at various phases of a product, process or service life cycle. With products in which the growth phase exposes serious technological deficiencies, the manager may be required to quickly develop a new technology within a very short time period. The need to insert a new or more robust or higher capability technology may also be driven by competitive forces. Enterprises are now competing against time, which requires reducing the time to develop, produce and deliver products, processes or services to the customer (Zirger and Hartley 1994). Rapid product development is a critical component of time-based competition.

During the technology's deployment within a product, process or service life cycle, it may also start to diffuse to other areas or products not associated with its original intended deployment. The technology can have societal and economic impacts which can change the behavior of society or the economy. At this final stage, the technology may have caused major socio-economic structural changes.

2.2 System Life Cycle

Many products and their technologies are embedded in complex systems, e.g., Boeing 777 aircraft. This further embedding of a technology further complicates the management process.

Blanchard and Fabrycky indicate that the system life cycle is characterized as a *consumer-to-consumer* process (Blanchard and Fabrycky 1990, p. 17). The *consumer-to-consumer* process consists of number of phases. This cyclical process is generic in nature and represents the life cycle of large-scale systems. The first phase of the process is the responsibility of the consumer and consists of identifying the system needs. In this identification phase, the consumer develops the needs due to obvious deficiencies or problems which have been determined through basic research.

The next group of phases are the responsibility of the system producer. Once the consumer has identified the system needs, the producer begins a system planning function. This function can consist of market analysis, development of feasibility studies, advanced system planning and development of a proposal for the consumer. The producer enters a system research function in which the various activities associated with research may occur, such as basic research, applied research which is needs oriented, development of research methods, and presenting the results of the research. These activities lead to a system design function by the producer:

- System planning function
- System research function
- System design function
- Production and/or construction function
- System evaluation function
- System use and logistic support function

Figure 5.2 shows this process, as concurrent sub-life cycles simplified to two basic phases, the acquisition and utilization phases. This total systems approach to life-cycle or concurrent engineering embraces the life cycle of the manufacturing process as well as the life cycle of the product service system.

2.3 Project Life Cycle

The project cycle is driven by the underlying cycle of the technology or system. This cycle can be characterized in terms of the resource allocation which has a very different level of requirements for human and fiscal resources. The project life cycle can be defined by the information requirements, which vary in amount and type of information needed at the various phases of the project. Finally, the project progress demonstrates a variation in terms of meeting milestones, performance goals and other measures of the progress in developing the product, process or service.

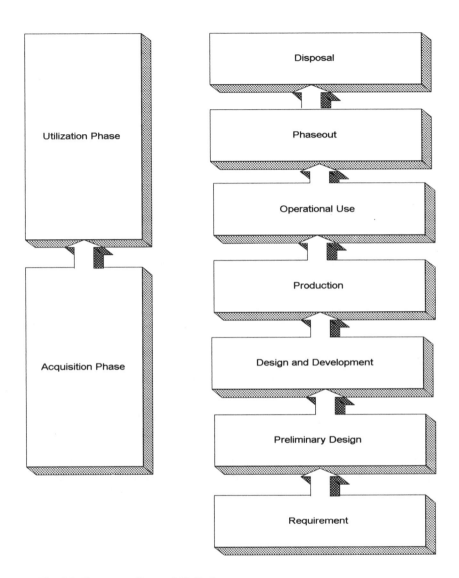

Fig. 5.2 Concurrent System Life Cycles

2.3.1 Resource Allocation

Figure 5.3 shows resource requirements over a project life cycle. In developing a new technology or system, the composition of the human, financial and other resources varies over the life cycle of the project. While Figure 5.3 is typical in terms of the cyclical nature of the distributions, the

154

actual maximums and declines would be different for the type of project which is managed and its complexity.

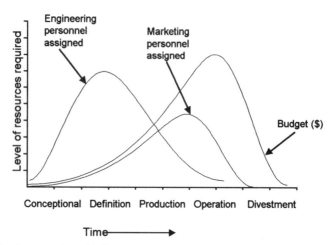

Fig. 5.3   Project Life Cycle Resource Requirements (Source: Cleland and King 1983, p. 249, © 1983 McGraw-Hill. Reprinted by permission of McGraw-Hill, Inc.)

2.3.2 Information Requirements

During the early stages of a project, an information system must be either developed or modified to meet the specific needs of the project.   This information system must acquire and process the specific project data, and be cognizant of the environment in which the project is embedded.   Forces in the environment, i.e., outside of the project organization, will also play an important role in determining the project's future.   Environmental information is, therefore, critical to effective project management.

2.3.3 Project Progress

According to Cleland and King (Cleland and King 1983, p. 248), there are three critical dimensions for assessing project progress during its life cycle. The resources being expended are one of the critical dimensions of project progress.   Maintaining a project within human and fiscal constraints allows an enterprise to meet its basic goals for a project.   Another critical dimension is timeliness of progress in terms of schedule and critical dates.   Independent of the cost effectiveness and timeliness, a project can fail if it is not meeting its performance goals.   The performance can be either in terms of physical requirements or in terms of market requirements, i.e., sales or market penetration.

2.3.4 Project Life Cycle Management

Cleland and King (Cleland and King 1983, p. 252-253) present a set of project management strategies associated with a five-stage private sector life cycle. This "Fox" (Fox 1973) life cycle is an alternative for presenting the project life. This life cycle, according to Fox, requires enterprises to exhibit flexibility. Many enterprises find it difficult to change as the project progresses through its cycle. This "Fox" life cycle consists of:

- Pre-commercialization
- Introduction
- Growth
- Maturity
- Decline

2.3.5 Project Life Cycle and Complexity

The project development cycle can be accelerated by reducing project complexity (Zirger and Hartley 1994). According to Zirger and Hartley, there are three dimensions of project complexity:

- Uncertainty
- Co-ordinative complexity
- Component complexity

Uncertainty

The absence of information contributes to the complexity of a new product, process or service development task due to the lack of technical and marketing information at the beginning of the project. The technical uncertainties arise at the outset of the development task as the technology is extended and changed. Market uncertainties arise from lack of adequate knowledge of the customer needs and finding the effective means to meet those needs.

Co-ordinative Complexity

The project life cycle is influenced by the amount of co-ordination required during the developmental cycle. The complexity increases non-linearly as the number of relationships increase among the required tasks. As co-ordinative complexity increases, greater functional co-ordination and integration requirements slow the process and increase the life cycle time. An interesting aspect of Zirger's and Hartley's (Zirger and Hartley 1994) research is that concurrent development can actually increase the co-ordinative complexity because more functional groups and activities require simultaneous co-ordination. However, while this may increase initial cycle time, if the concurrent process is successful, this loss in cycle time may be recaptured downstream because fewer delays occur due to engineering or process changes.

Component Complexity

Component complexity is defined as the number of distinct acts that need to be executed in the performance of the task. The number of distinct information cues that must be processed in the performance of those acts also results in component complexity. Incremental project change, part reduction, freezing the design early and vendor management can reduce component complexity.

2.4   Organizational Life Cycle

Organizations exhibit life cycles as they evolve, and can be divided into the five phases (Cleland and King 1983, p. 237)[3]. These stages are critical to the managers of technology since what is appropriate for a newly formed enterprise may not be appropriate for a mature enterprise, but could form a barrier to bringing a new technology to fruition. In the *creativity* phase, emphasis is placed on creating a new technology and market for the emergent enterprise. While in the *directional* phase, the enterprise is in a period of sustained growth.

The enterprise evolves from the successful application of a decentralized enterprise structure in the *delegation* phase. The *co-ordination* phase is characterized by the use of formal systems for achieving greater co-ordination within the enterprise. Finally, the *collaborative* phase is a crisis stage in the enterprise's life cycle, where a strong management team approach is required to overcome overly complex management system "red-tape". This final phase, if successful, includes solving problems through team action,

---

[3] Each of the five phases is given the name of the activity which predominates during that phase (Cleland and King 1983, p. 237). In fact, all activities may occur to some extent in all of the phases.

multifunctional teams, and frequent use of matrix team formation, i.e., targeting teams to problems.

The system of organizational life cycle shown in Table 5.1 implies that successful organizations go through a life cycle where small groups grow larger and more complex until a "crisis" requires re-engineering of the organization into a small, collaborative team, which is capable of dealing with complexity and size issues.

2.5   Business Life Cycle

Technologies and enterprises tend to evolve in a consistent pattern.   The business cycle, according to Howard, Jr. and Guile (Howard Jr. and Guile 1992, p. 11-14) can be divided into emergence, diffusion, development, and maturity.

2.5.1 Emergence (Market-driven opportunities)

The emergence phase of the business cycle is when a new technology is pushing into a market, or the market is pulling an existing technology into new applications.   New products, processes, or services that arise from a perceived market opportunity are called technology-driven opportunities. The challenge for the technology manager in these market-pull situations is in the search for the best technology, either internal or external to the enterprise. The manager is also challenged by the process of mutual adaptation between market and technology as an enterprise uses the technology to meet untapped market demand and create a business.

2.5.2 Diffusion   and   Development   (Product   and   process/service improvement opportunities)

The rise of a dominant design, process, or application increases the pace of diffusion and changes the nature of competition.   In this phase, competitive factors become very important. This process causes price competition to become more important, i.e., less product characteristics differentiation. This is the stage that the personal computer ("PC") market reached as it moved into a commodity phase.   This change had serious impact for enterprises which found their PC dominant market share overtaken by lower-priced PCs.

2.5.3 Maturity (End-game opportunities)

In the maturity phase, the pace of technological change slows down for a particular application, product, process or service, opening the way for a new innovation. Businesses face uncertainty about the values and risks inherent in further investment into existing and mature groups of products, processes, services or applications. For example the introduction of high-speed jet container vessels may cause a major change in the marine shipping field.

2.6 Life Cycle Interactions

Technology life cycles can have significant impacts on society. The following scenario is a simplified case of the interaction of technology, productivity and worker requirements. This scenario is presented to illustrate that life cycles can have unexpected results in the socio-economic arena.

A new technology is introduced into an industry; it encounters initial resistance due to a number of factors. Since the technology has a definite productivity value, i.e., it allows workers to produce more units in a similar time, it is eventually established. As the technology makes further penetration into the same industry, more enterprises employ the technology. Depending on the growth of the industry and decreased cost, the number of workers required will either remain the same producing more, or decrease to further decrease cost, or increase. History indicates that the number of the original workers will then decrease ("right-sizing"), and the remaining workers will be replaced by a smaller number of workers with the newer technological skills.

An example of "right-sizing" is the introduction of electronic type setting and the eventual phasing out of type setters who were replaced by a smaller number of computer operators and programmers. The following is a simplified model of this situation using a S-shaped growth model based on the Pearl equation.

If we assign:

u = units in use at time t
U = upper limit of units which can be placed in use
t = time
$a_u b_u$ = coefficients for units

Then:

$$u = \frac{U}{1 + a_u e^{-b_u t}}$$

If the productivity of workers is a function of the number of new technologies in use displaced by a learning factor, assuming perfect learning and p is the productivity at time t and is given by a productivity factor A

$$p = A u = \frac{A U}{1 + a_u e^{-b_u t}}$$

If we assume the requirement for units in an industry using the technology is r at time t and is also given by a Pearl type equation, then:

$$r = \frac{R}{1 + a_r e^{-b_r t}}$$

where R is the limit of the requirements and a_r and b_r are coefficients for the requirements equation. Then the number of worker (w) required is given by:

$$w = \frac{r}{p} = \left(\frac{R}{A U} \right) \left(\frac{1 + a_u e^{-b_u t}}{1 + a_r e^{-b_r t}} \right)$$

Then for a constant requirement r $a_r = 0$ and the number of workers working in industry using the technology is represented by:

$$w = \left(\frac{R}{A U} \right) \left(1 + a_u e^{-b_u t} \right)$$

Figure 5.4 shows a resulting reverse S-shaped curve. If the requirements follow an S-shaped curve we find a situation depending on the constant factors where the number of workers can either decrease or increase over time. History indicates a trend which consists of a cyclical curve around an increasing slope as new technologies stimulate new users, also increasing productivity. Figure 5.5 shows the cyclic motion and the fact that the worker levels increase and then may have to decrease as the new productivity reduces the need for workers.

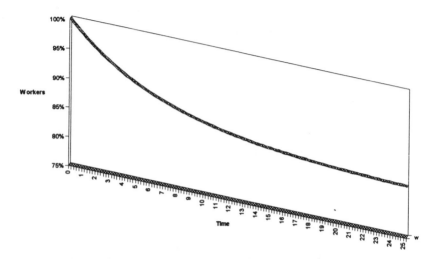

Fig. 5.4 Relation Between Technology and Worker Requirements

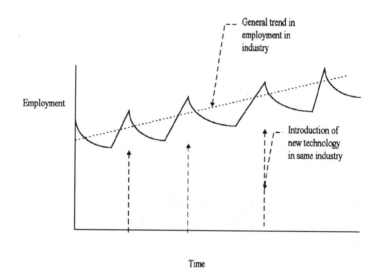

Fig. 5.5 Potential Impact of a New Technology in an Industry

3.   DECISION MAKING

Because the future is unpredictable, the management of technology is intermediately associated with decision making and risk assessment in the presence of uncertainty. The dictionary defines the noun *decision* as (Stein 1966):

> *"1. the act of deciding; determination, as of a*
> *question or doubt, by making a judgment."*

The decisions a manager of technology must consider are project selection, financing, budgetary issues, internal versus external acquisition, and project termination. Due to the uncertainty of future actions of competitors, availability of resources, markets, and other potential internal and external environmental variables, the ability to assess these uncertainties and associated risks are a critical element in the total technological process. Figure 5.6 shows a decision analysis flow chart according to Clemen (Clemen 1991, p. 6).

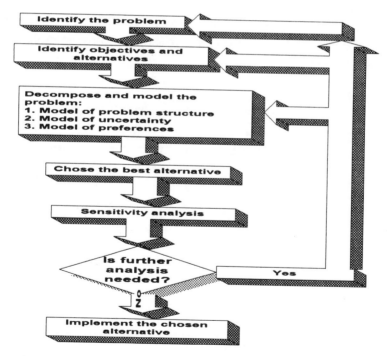

Fig. 5.6   Decision Flow Analysis According to Clemen (Source: Clemen 1991, p. 6, © 1991 Wadsworth Publishing Co. Reprinted by permission.)

According to Martino (Martino 1993, p. 251-252), technological decision making involves a number of steps. The process is initiated by an action, i.e., either to change the present situation or a deliberate decision not to change. The next step is to determine feasible courses of action and if there are no feasible courses of action or only one, then there is no decision. If there are a multiplicity of courses of action, i.e., more than one are available, then a decision problem exists. A multiplicity of courses of action requires the decision maker to make a selection, i.e., choosing from available courses of action usually under a system of constraints. Finally, once a course of action is decided it requires implementation, i.e., doing something. It must be also remembered that not doing something is also a decision.

Personal judgments about uncertainty and values are important inputs for decision analysis. In any enterprise, there are three levels of decisions according to Athey (Athey 1982, p. 49). Strategic decisions determine overall system objectives. Tactical decisions assess how to best accomplish the overall system objectives. Finally, operational decisions require the implementation of system objectives, while keeping the system within constraint limits. Table 5.2 shows the relationship between problems and these decision levels.

Table 5.2
Relationship Between Problems and Decision Levels

Type of Problem	Decision Level	Type	Activity
Negative	Operational	Maintenance	Preservation Safekeeping Keeping it alive
Positive	Tactical	Management Planning	How to make it more effective Better benefit/cost ratios
Fundamental	Strategic	Policy	Direction Technological vector

(Based on data contained in Athey 1982, p. 50)

3.1   Decision Process

The decision making process should be based on logic, considering all available data, possible alternatives and the systems approach. This process consists of defining the problem and the factors that have an influence, establishing the decision criteria and goals. A decision model is formulated and used for the identification and evaluation of alternatives. Using this model and criteria, an alternative is chosen and this decision is then implemented.

It is very important that a clear, concise statement of the problem is developed, the most critical part of the process, and a very difficult step in practice. According to Heizer and Render, there are major factors which make stating the problem a difficult step (Heizer and Render 1991, p. 57). One of these factors is the conflicting viewpoints of the participants. The impact of the problem on various portions of the enterprise may also make it difficult to accurately develop a statement of the problem. Validity of initial assumptions and the tendency to formulate the problem, in terms of solutions, can make it very difficult to state the problem in a realistic manner. Finally, the technological change can obviate a problem before it is even solved. The collapsing time scale for technological developments can make the former decision problem a non-problem.

Once the problem formulation stage has been completed and goals have been clearly defined, the next step is the development of a model which can be used to assess the available alternatives. Model formulation can take many forms. The model, due to the complexity of the real situation, can only be a representation. It is important when formulating the model to make sure it truly represents the problem. The manger of technology must understand the limitations of the model, and its structure. Validating all data used in the model will help the technology manager in reaching a valid decision option.

The technology manager must be fully aware of the advantages and disadvantages of models. According to Heizer and Render (Heizer and Render 1991, p. 63), models are developed because they are less expensive and disruptive than experimenting with the real world. However, it must be considered that there have been numerous instances where a "Gordian Knot"[4]

---

[4] Gordian knot ,1 : an intricate problem or a problem insoluble in its own terms 2 : a knot tied by Gordius, king of Phrygia, held to be capable of being untied only by the future ruler of Asia, and cut by Alexander the Great with his sword.

164

approach would have been more cost effective than development of a detailed and costly model.

Models, however, are amenable to techniques to check sensitivity, a difficult thing to accomplish in the real world. The development of a model by a technology manager can encourage input from all members associated with decisions by providing a framework for discussion. A model forces consistency and a systematic approach which can reduce error. Models require the technology manager to use specificity in developing constraints and goals. Without this specificity, a model is like a ship without a rudder in a storm. If the model is an accurate portrayal of the problem, then its use can reduce decision making cycle time.

Disadvantages, according to Heizer and Render (Heizer and Render 1991, p. 63), include the fact that models can be expensive and time-consuming to develop and test. Furthermore, models are open to misuse and misinterpretation, especially if a decision maker is biased toward a particular outcome. While models are quantitative, it should not be forgotten that qualitative information also has a role in the decision making process, a fact sometimes overlooked. It must be re-emphasized that a model is only a representation of the actual decision problem.

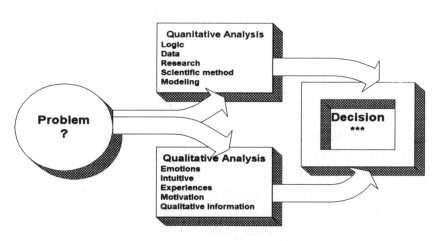

Fig. 5.7 Quantitative and Qualitative Decision Making

Alternatives for evaluation are the set of possible courses of action which encompasses all the feasible solutions to the problem. It is very important to identify as many members of this set of potential solutions as possible. This stage of the process, i.e. identification of alternatives, requires in most instances creativity, since if the solution was obvious, there would likely not be a problem. Figure 5.7 shows both the quantitative and qualitative method of arriving at a decision from the set of feasible solutions.

3.2   Information Acquisition

Decision making requires information. Figure 5.8 shows a framework for information acquisition based on the media used for transmission. Media refers to the various mechanisms used to transfer information from a source to the decision maker (Jones et al. 1994). An information link consists of the source of the information, the contact, or nature of information, and the medium that is used to communicate the information. This information link can occur at any point in the decision process. Decision context, source, media and decision making phase are critical to the framework shown in Figure 5.8. Decision making models usually do not deal with what is the appropriate media used for information gathering while making decisions.

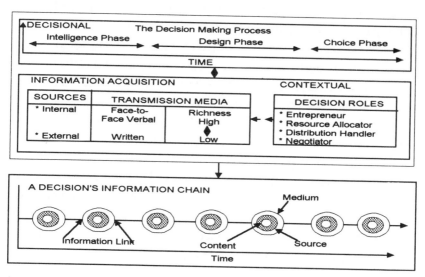

Fig. 5.8 Information Acquisition Framework (Source: Jones et al. 1994, © 1994 IEEE. Reprinted by permission.)

A decision's information chain is a time-ordered sequence of information links associated with a specific decision. The rate information links are added to a chain per period of time is defined as the chain's *velocity*. Managers of technology receive information more frequently in high velocity chains than in low-velocity chains. A decision in a high-velocity chain may be fast-paced because of time constraints placed upon it. The manager of technology as a receiver in these high-velocity information chains must determine how to respond to each link in a relatively short period of time (Jones et al. 1994).

3.3 Roles of the Technology Manager

The manager of technology can have several roles in the decision process, according to Jones et al. (Jones et al. 1994). As an *entrepreneur* the technology manager acts as an initiator and designer of controlled technological change. The manager in this role acquires information using a relatively high percentage of information-lean media[5], i.e. reports, brochures, and periodicals. The entrepreneur has an idea creating role. The manager of technology is a *resources allocator*, working with internal enterprise information sources when ascertaining the resource asset base and its appropriate distribution. Managers of technology rely on information-lean media to reduce the uncertainty in overseeing the process. This information includes past budgets, formal procedures and written documents. As a *disturbance handler*, the manager of technology, according to Jones et al. (Jones et al. 1994), deals with situations and changes that are partially beyond their control when the pressures on the enterprise are too large to be ignored. These disturbances or problems arise very suddenly and require immediate action. Under these conditions, the manager of technology is less likely to use written media and scheduled meetings than verbal or face-to-face meetings when trying to resolve problems. Handling disturbances may require rich[6] media, i.e., meetings with particular individuals and multiple channels. Finally, the manager is a *negotiator* where he or she enters negotiation with enterprises and individuals on behalf of the unit. Extensive face-to-face scheduled interactions are to be expected in negotiating activities. This role demands rich media, such as face-to-face meetings.

3.4 Decision Making Tools

There are many tools available to the manager of technology to assist in the decision making process. These tools can be classified as either financial

---

[5] Information lean media consists primarily of printed and electronic information.

[6] Richness refers to the amount of face-to-face and verbal means of transferring the information. The richest form is face-to-face communications.

decision tools or general decision support tools. The basic financial decision tools consist of life-cycle cost analysis, cash-flow analysis, and cost-effectiveness analysis. The decision support tools include decision trees, influence diagrams, systematic hierarchical models, and fuzzy logic models.

3.4.1 Financial Decision Tools

Many of the decisions that a technology manager faces involve financial questions. There are a number of methods to use when making financial decisions. These can be as simple as a profit and loss statement, which is a calculation model of a system which can be used for predictive purposes. This type of model is limited as to the predictions which can be made from it. However, qualitative predictions can be made more easily than they could be made without it (Cleland and King 1983, p. 97-99).

Life Cycle Analysis

Technological decisions, particularly in the early stages of the life cycle, have life cycle implications and affect life cycle costs (Blanchard and Fabrycky 1990, p. 501). Figure 5.9 shows a life cycle cost model for an electric power station. Figure 5.10 shows sample life cycle cost analysis profiles.

These cost profiles, according to Blanchard and Fabrycky (Blanchard and Fabrycky 1990, p. 523-524) can be developed by identifying all activities throughout the life cycle that will generate costs of one type or another. This can include the cost of the externalities such as environmental costs. The manager of technology will have to relate each activity to a specific cost category in the cost breakdown structure. Establishing appropriate cost factors in constant dollars[7] for each activity in the cost breakdown structure is another critical step in the life cycle cost process. This requires projecting individual cost elements into the future on a year-to-year basis over the life cycle and using appropriate inflationary factors, economic effects of learning curves, and changes in price levels. Finally, the process requires summarizing individual cost streams by major categories and developing top-level cost profiles. If this process is followed, then various alternatives can be compared on a total life cycle basis, as shown in the example in Figure 5.11.

---

[7] Constant dollars are dollars of equal buying power and are expressed in terms of base year dollars.

168

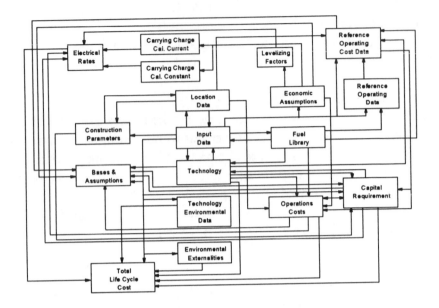

Fig. 5.9 Life Cycle Cost Model for Electric Power Station

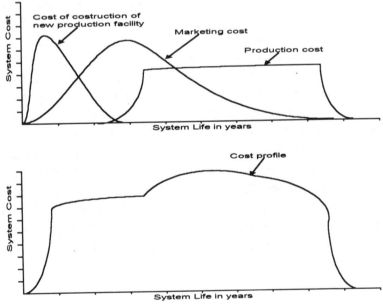

Fig. 5.10 Sample Life Cycle Cost Profiles

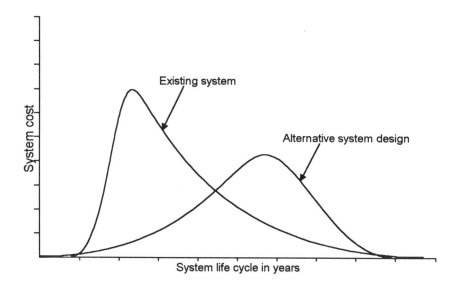

Fig. 5.11 Comparison of Systems Based on Life Cycle

Cash-Flow Analysis

The life cycle cost analysis can serve as an input into a cash-flow analysis. The cash-flow analysis allows a manager to determine when a project will start to recoup investment and overall return on investments over the lifetime of the technology.

Figure 5.12 presents a model of flow of cash through an enterprise according to Cleland and King (Cleland and King 1983, p. 135). The results of a cash-flow analysis can be presented in terms of a break-even or payback chart.

Time to break-even analysis is a simple technological decision management tool which can be used throughout the course of a technology project (Howard Jr. and Guile 1992, p. 66-69). *Time to break-even analysis* summarizes technology development experience, current outlook and entrant goal. Figure 5.13 shows a cash-flow versus time analysis. Early project development costs are not returned until income starts to materialize after the product, process or service introduction. The time to break-even is when the investment reaches zero. This type of analysis can also be used to track project life cycle cost.

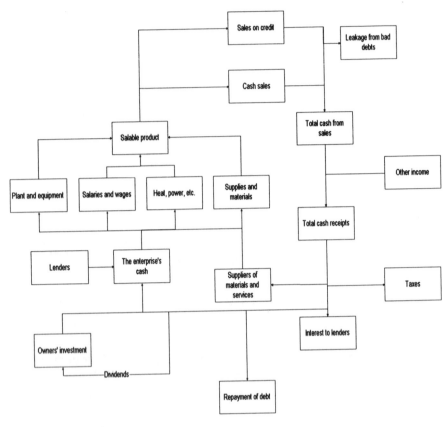

Fig. 5.12 Model of Flow of Cash Through an Enterprise (Source: Cleland and King 1983, p. 135, © 1983 McGraw-Hill. Reprinted by permission of McGraw-Hill.)

Another financial analysis technique which can be used is cash-flow ratios (Giacomino and Mielke 1995). While Giacomino and Mielke applied this approach to a total enterprise, it may be possible to apply it also to a particular technological development embedded in an enterprise. Financial analysts use these ratios to predict financial variables to evaluate relative performance. The technique uses a relative performance evaluation approach by comparing performance to a chosen industry or benchmark ratio. The ratios that Giacomino and Mielke used in their study for enterprises are shown in Table 5.3. While these ratios are for total enterprises or large sub-units, it may be possible to formulate a *virtual project enterprise* where a modified set of these ratios can be employed.

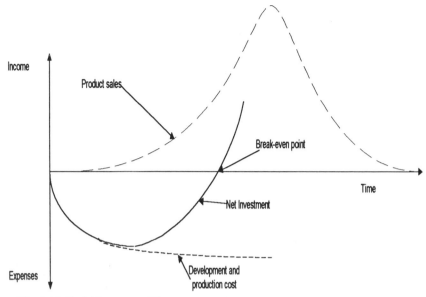

Fig. 5.13 Cash Flow versus Time

Table 5.3
Cash Flow Ratios

Sufficiency Ratios	
Cash flow adequacy	Cash from operations/Long-term debt paid + purchases of assets + dividends paid
Long-term debt payment	Long-term debt payments/Cash from operations
Dividend payout	Dividends/Cash from operations
Reinvestment	Purchase of assets/Cash from operations
Debt coverage	Total debt/Cash from operations
Depreciation-amortization impact	Depreciation + amortization/Cash from operations
Efficiency Ratios	
Cash flow to sales	Cash from operations/Sales
Operations index	Cash from operations/Income from continuing operations
Cash flow return on assets	Cash from operations/Total assets

(Source: Giacomino and Mielke 1995, © 1995 IEEE. Reprinted by permission.)

Cost Effectiveness Analysis

Another tool for technology decision making is cost-effectiveness analysis. The objective of this analysis is to determine the relative merits and difficulties of a particular course of action. This methodology measures either current development cost, life cycle cost, or some other measure of cost against effectiveness or benefits. This form of decision making is one of the favorite methods used by the governmental sector in deciding between various alternatives.

3.4.2 Decision Support Tools

Financial analysis tools serve as sub-units of a decision support system ("DSS"). A DSS can be used in conjunction with a systems approach to solve a number of decision problems. A DSS can also be aimed at each step of the decision process.

Decision Trees

Decision trees are a class of evaluation models designed to incorporate the notion of risk into the decision process. Figure 5.14 shows an example of a decision tree for the introduction of new technological processes. Decision trees consist of four basic elements: decision nodes, chance nodes, probabilities, and payoffs. *Decision nodes* indicate all the possible courses of action available to a decision maker. *Chance nodes* show the intervening uncertain events and all of the possible outcomes. Each chance node has an associated *probability*, i.e., an estimate the possibility of a chance event occurring. Finally, *payoffs* summarize the consequences of each possible combination of choice and chance.

Decision trees are particularly effective in structuring and visualizing complex decision problems under uncertainty. In contrast with deterministic models such as linear programming, these tools are stochastic in nature. They incorporate probability and the concept of sequential nodes that form different outcome paths. The concept of expected value is employed to aggregate the expected benefits or costs of a particular course of action with the probability of it occurring.

Decision trees are appropriate for decisions where there are a limited number of mutually exclusive alternatives with an identifiable set of criteria for selecting among alternatives. They have been applied to a variety of decision problems and investment decisions, such as oil drilling, location decisions, and R&D portfolio selection.

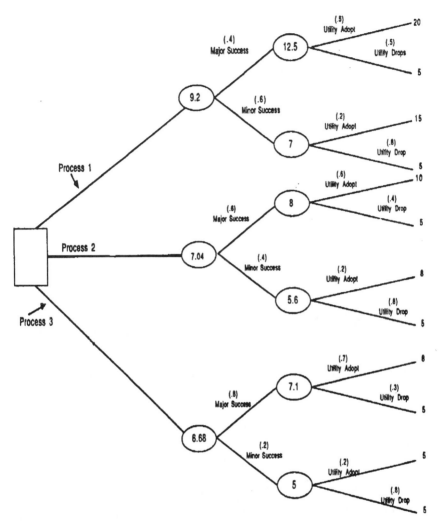

Fig. 5.14 Example of a Decision Tree for the Introduction of New Technologies by a Utility

Influence Diagrams

An influence diagram is a graphical representation of a decision problem. A decision problem consists of decisions to make, uncertain events, and values of outcomes. An influence diagram uses various geometric shapes to represent these elements of the decision problem. The shapes are linked with arrows in specific ways to show the relationships among the elements.

Figure 5.15 shows an example of a simple influence diagram. Solid arrows between circles represent that one outcome is relevant for assessing the chance associated with the other event, i.e., in Figure 5.15 the number of new processes licensed is impacted by the potential competitive processes which may enter the market in the future. A rectangular event is a decision event. Decision events connected to a chance event imply that the decision is relevant for assessing the associated chance event, e.g., in Figure 5.15 the price for the new process license is relevant to the number of new processes which are licensed. A dashed line between decision events implies that one decision is made before the other decision event. Influence diagrams are excellent for showing the structure of a decision, but many details are hidden. Decision trees are used to represent the same decision process but with greater detail. Influence diagrams can be mapped into decision trees as is shown by Clemen (Clemen 1991).

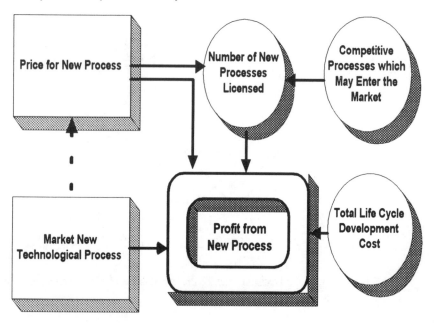

Fig. 5.15 Example of Influence Diagram

Systematic Hierarchical Models

Strategic R&D decisions are part of a complex problem, which involves (Liao et al. 1995) formulation of technological strategy (see Chapter Two - *Technological Strategy*), determination of competitive strategies strategy (see

Chapter Two - *Technological Strategy*), evaluation of R&D strategies (see Chapter Two - *Technological Strategy* and Chapter Three - *Technological Forecasting*), and selection of R&D projects (see Chapter Four - *Generation of Technology*). These phases are hierarchically related and sequentially interconnected according to Liao et al. (Liao et al. 1995). Figure 5.16 shows the form of the Liao et al. model. The main features of this model include systematic integration., chain effects, and interactions with the environment.

Systematic integration is an important element in modeling the strategic R&D decision making process. Consistency must be maintained throughout this process. The process can be considered in terms of chain effects in which the choice in one module of the integration process determines the choice in subsequent modules. This type of decision model supplies a consistent framework for strategic R&D decision making by systematically integrating the associated decision making tasks. Systematic integration emphasizes the assessment of environmental effects relevant to each decision task, and considers multiple criteria for decision making. This enables decision makers and leadership teams to actively respond to environmental changes by adjusting criteria and decisions on a general equilibrium basis. The process results in decisions which are internally consistent.

Fuzzy Logic

Fuzzy logic is a part of artificial intelligence which has significantly enhanced knowledge-based or expert systems technology (Dutta 1993). The objective of fuzzy logic is to model the imprecise modes of reasoning that have an important role in making rational decisions in an environment of uncertainty and imprecision. Fuzzy logic is used in correcting an important deficiency of conventional knowledge-based systems by dealing with vague data and modeling of imprecise reasoning procedures. The integration of expert system concepts and fuzzy logic has led to a more robust modeling of uncertainty and imprecision.

It is possible to use the fuzzy logic methodology to formulate models of technological portfolio selection and other decisions which face managers of technology. In many instances, a manager of technology is faced with decisions which have imprecise information inputs. A fuzzy logic process integrated with an expert system with embedded enterprise experience may prove more robust than other methods in the uncertain environment of technological development.

176

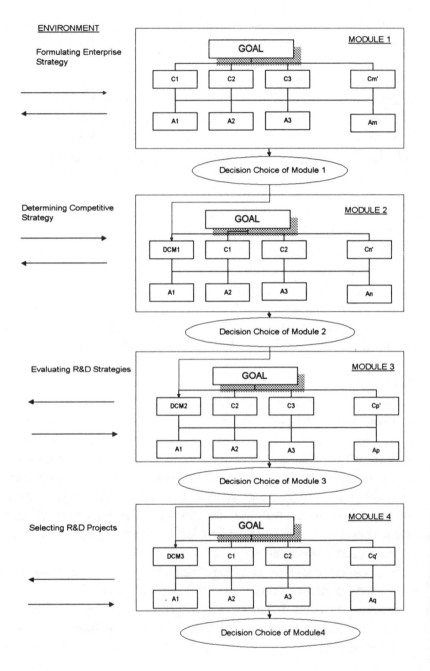

Fig. 5.16 Systematic Hierarchical Models (Source: Liao et al. 1995, © 1995 IEEE. Reprinted by permission)

4. UNCERTAINTY AND RISK ANALYSIS

4.1 Uncertainty

Uncertainty of the future state of the internal and external environments is the principal reason why it is difficult to make decisions. Uncertainty can arise because of a number of factors (Morgan and Henrion 1990, p. 47-72). Incomplete information is a major source of uncertainty. The reduced information set is one of the major causes of uncertainty. Due to the nature of the universe, it is highly improbable that a manager of technology will ever have a full set of the information necessary for decision making.

Another source of uncertainty is disagreement between information sources. Linguistic imprecision adds to uncertainty, since language can only convey a portion of the total meaning. Natural variability, unknown quantities and imprecise structure of a decision model cause uncertainty.

In decision models, according to Morgan and Henrion (Morgan and Henrion 1990, p. 56-67) uncertainty usually arises due to uncertainty in empirical quantities, which generally constitute the majority of quantities in risk analysis. The sources of uncertainties in empirical quantities include: statistical variation, subjective judgment, linguistic imprecision, variability, inherent randomness, disagreement, and approximation. It is important to include uncertainty explicitly when the attitude of the decision maker toward risk is important, and uncertain information from various sources must be combined.

There are various analytic and computational techniques for examining uncertainty. *Sensitivity analysis* is one of the approaches, where computation of the effect of changes in inputs in model predictions are analyzed. Another technique is *uncertainty propagation*, of which is the determination of uncertainty in the decision model outputs induced by uncertainties in its inputs. Another technique, *uncertainty analysis*, compares the importance of the input uncertainties in terms of their relative contributions to uncertainty in the outputs.

4.2 Risk Analysis

Due to numerous unknowns, new technological products, processes and services are accompanied by substantial risk. The definition of risk involves
"exposure to a chance of injury or loss".

Due to *chance*, risks may be perceived and not actual (Wharton 1992). Managers of technology make decisions based on actual risks and perceptions about the consequences of their actions. Whether real or imagined, however, perceived risks have to be taken into account in decision making.

The evaluation and comparison of economic risks usually takes the form of a cost-benefit analysis. R&D portfolio analysis is basically minimizing the financial risk of losing a substantial amount while insuring a satisfactory return by choosing a diversified portfolio of R&D projects. One approach for security portfolio analysis is that designated the *Markowitz* model (Thomas 1992). This model can be restated for use in a technological portfolio analysis, and assumes the maximization of the expected utility of the R&D portfolio over a single period. A logarithmic utility function implies that this criterion makes sense for multi-period problems. The expected return and variance of each R&D project, over the period, is known. The variance between the projects is another assumption used in this model. The R&D portfolio, consisting of a combination of projects, can be described from the manager's viewpoint by the mean and variance of its expected returns. The manager's utility function depends only on these assumptions and is risk averse.

Where

$$x_i = \frac{\text{value of funds invested in project i}}{\text{total value of funds invested in R\&D portfolio}}$$

the expected return of the R&D portfolio is

$$R_p = \sum_i x_i r_i$$

while the variance of the expected returns on the R&D portfolio is

$$V_p = \sum_i x_i^2 \sigma_i^2 + \sum_i \sum_j x_i x_j \sigma_{i,j}^2$$

where

$$\sum_i x_i = 1$$

Most R&D managers' utility functions, u, are functions of R and V, i.e. $u(R,V)$ and these functions are monotonic increasing of R and decreasing in V. Thomas' (Thomas 1992) security approach is used to determine the Pareto[8]

---

[8] Each portfolio is presented as a bar graph, and the bars are then arranged side-by-side in descending order, beginning with the item with the highest value and ending with the lowest-valued portfolio.

optimal R&D portfolios, i.e., those which are not dominated by other portfolios. Thus Portfolio 1 with mean R_1 and variance V_1 dominates Portfolio 2, with mean R_2 and variance V_2 if $R_1 \geq R_2$, $V_1 \leq V_2$ and at least one inequality is strict. Pareto optimal portfolios are those that minimize $-\beta R_p + V_p$ for some $\beta > 0$. Finding Pareto optimal portfolios reduces to solving the following as β varies.

Minimize

$$-\beta \sum_i x_i r_i + \sum_i \sum_j x_i x_j \sigma_{i,j}^2 + \sum_i x_i^2 \sigma_i^2$$

subject to

$$\sum_i x_i = 1$$

$$x_i \geq 0 \qquad i = 1, 2, \ldots \ldots, n$$

4.2.1 Technological Risk Management

The management of technological risk is best accomplished by the maximization of expected values, avoidance of catastrophes, and ignoring remote possibilities (Wharton 1992). Technology project risks increase with the level of uncertainty, the probability of failure, and the required financial commitment. There are various types of risks (Garvin 1992, 329-330) that a manager of technology must manage include: *market risks*, i.e., demand-side uncertainties, and *technological risks* consisting of unknowns of newly developed materials, processes and concepts. *Competitive risks,* or difficulty of predicting other enterprises' responses are another type of risk which must be considered. *Organizational risks* are the internal risks due to an inability of traditional structures, staffing and reporting relationships to accommodate new technologies. The manager of technology must consider *production risks*, i.e., the difficulties of start-up and implementation. The final risk which must be considered is *financial risks* which arise due to the large investments placed at risk with uncertain payoffs. These risks are interrelated, thus adding to the complexity of the decision process and increasing opportunity for chaotic behavior. However, it is important to distinguish among these risks because each imposes its own demands on the manager of technology.

After risks have been identified and a desired *risk profile* is selected, the next step is risk reduction. As resulting action causes movement away from the existing state in any of the risk multi-dimensional phase space, i.e., market, competitive, technology, organization, production or financial, the larger the risk becomes. However, because of inter-relationships no system remains

180

static, and stochastic factors also will require movement and thus increased risk.

4.2.2 Attitude Toward Risk

Managerial attitudes toward risk exert considerable influence on technological decision making. An understanding of the manager of technology's attitude toward risk is important because decisions are very different depending on his/her willingness to take risks. These attitudes influence the range of considered choices (Pearce and Robinson 1991, p. 284). As the attitude toward risk increases, i.e., favors higher risk, the range of alternatives expands. Where the attitude of the manager of technology is risk averse, the range of choices becomes more limited as risky alternatives are eliminated. This implies that it is very important to provide tools which afford the manager the ability to more accurately portray risks and a means of mitigating them.

REFERENCES

Athey, T. H. (1982). *Systematic Systems Approach*, Prentice Hall, Englewood Cliffs, NJ.

Blanchard, B. S., and Fabrycky, W. J. (1990). *Systems Engineering and Analysis*, Prentice Hall, Englewood Cliffs, NJ.

Cleland, D. I., and King, W. R. (1983). *Systems Analysis and Project Management*, McGraw-Hill, New York, NY.

Clemen, R. T. (1991). *Making Hard Decisions*, Duxbury Press, Belmont, CA.

Dutta, S. (1993). "Fuzzy Logic Applications: Technological and Strategic Issues." *IEEE Transactions in Engineering Management*, 40(August), 237 - 254.

Fox, H. W. (1973). "A Framework for Functional Co-ordination." *Atlantic Economic Review*, 23, 10-11.

Garvin, D. A. (1992). *Operation Strategy: Text and Cases*, Prentice Hall, Englewood Cliffs, NJ.

Giacomino, D. E., and Mielke, D. E. (1995). "Cash Flows: Another Approach to Ratio Analysis." *IEEE Engineering Management Review*, 23(Spring), 56 - 59.

Heizer, J., and Render, B. (1991). *Production and Operations Management: Strategies and Tactics*, Allyn and Bacon, Boston, MA.

Howard Jr., W. G., and Guile, B. R. (1992). *Profiting from Innovation: The Report of the Three-Year Study from the National Academy of Engineering*, The Free Press, New York, NY.

Jones, J. W., Saunders, C., and McLeod Jr., R. (1994). "Information Acquisition During Decision Making Processes: An Exploratory Study of Decision Roles in Media Selection." *IEEE Transactions in Engineering Management*, 41(February), 41 - 49.

Liao, Z., Greenfield, P., and Cheung, M. T. "An Interactive Systematic Hierarchy Model for Strategic R&D Decision Making in a Dynamic Environment." *IEEE Annual International Engineering Management Conference*, Singapore, 321 - 326.

Martino, J. P. (1993). *Technological Forecasting for Decision Making*, McGraw-Hill, Inc., New York, NY

Morgan, M. Granger, and Henrion, Max, Cambridge, UK,. (1990). *Uncertainty: A Guide to Dealing with Uncertainty in Quantitative Risk and Policy Analysis*, Cambridge University Press, Cambridge, UK.

Pearce, J. A., II , and Robinson , R. B., Jr. (1991). *Strategic Management: Formulation. Implementation, and Control*, Richard D. Irwin, Inc., Homewood, IL.

Stein, J. (1966). *The Random House Dictionary of the English Language*, Random House, New York, NY.

Thomas, L. C. (1992). "Financial Risk Management Models." Risk: Analysis Assessment and Management, J. Ansell and F. Wharton, eds., John Wiley & Sons, Chichester, UK, 53-70.

Wharton, F. (1992). "Risk Management: Basic Concepts and General Principles." Risk: Analysis Assessment and Management, J. Ansell and F. Wharton, eds., John Wiley & Sons, Chichester, UK, 1-14.

Zirger, B. J., and Hartley, J. (1994). "A conceptual model of product development cycle." *Journal of Engineering and Technology Management*, 11(December), 224 - 251.

DISCUSSION QUESTIONS

1. What are some of the key factors to consider in viewing the technology life cycle? What are some of the key decision points involved during a technology's life cycle?

2. Each system goes through a life cycle. Take a system of your choosing and trace the stages of its life cycle and some of the key questions that should be asked about the system during its life cycle.

3. Comment upon the life cycle of xerography and how it impacted productivity and the need for workers.

4. Choose a technology and comment upon the impact on an enterprise or market of this technology and where is this technology in the life cycle, and what are the upcoming decision points?

5. Why should a manager of technology consider uncertainty?

6. What are the uncertainties in the development of new software products?

7. What are the decisions facing the manager of technology in:
   Small entrepreneurial firm.
   Large governmental laboratory.
   Government agency such as NASA, DOD, etc.
   Large multinational enterprise.

Foreign owned technology enterprise operating in the United States.

What are the major similarities, differences, and other factors that distinguish these decisions?

8. Choose a technology and prepare a qualitative risk analysis for this technology.

CHAPTER 6

Enterprise Structure and Design

1. INTRODUCTION

Technology develops within an enterprise structure. The effectiveness of not only the enterprise, but the individual technological developments are directly related to the type, style and form of the enterprise. The enterprise structure combined with the processes used within the structure form the basis for translating a technological innovation from conception to introduction into the environment in which it is used.

One of the major tasks in managing technology is to design the enterprise structure. This structure provides the framework for translating technological innovations into products, processes and services that are accepted and used within the intended environment. Classical enterprise structuring or organizing, consists of determining what activities need to be accomplished, grouping the these activities in terms of a functionally oriented structure of some type and staffing the structure with appropriate skilled human resources to perform the designated activities in a co-ordinated manner (Blanchard and Fabrycky 1990, p. 556-561).

The introduction of new technology can seriously impact including significant socio-economic conditions. The internal factors can strongly influence the process of developing a new technological product, process or service. The manager of technology should have an understanding of these factors and be aware of tools for assessing and mitigating impacts. Enterprise structures reflect relationships, while the processes govern the activities that take place within these structures. Taken together they represent the *technological delivery system* (Porter et al. 1991, p. 308).

Enterprises that are involved in technology perform four functions: research and development, production and operations, marketing, and administration. The initiator of the technological process is *research and development.* The *production and operations* function translates the technological innovation into a product, process or service which can be used in an intended environment. *Marketing* generates the demand or serves as the translator of customer need into an enterprise request or an order. The *administration* function provides the resources, monitors progress, and provides financial services.

Table 6.1 shows a comparison of how these functions are exercised in various enterprises. These functions are performed to various degrees by religious and academic institutions, government organizations and business enterprises.

2. ENTERPRISE STRUCTURES

The structure and adaptability of an enterprise are the basis for organizational success and survival. The enterprise processes identify key activities and the manner in which they interact to achieve the enterprise's objectives and goals. The structure of an enterprise is composed of its major organizational elements and their relationships. The basic structures around which an enterprise is designed include (Pearce and Robinson 1991, p. 327-335):

- Functional
- Geographic
- Divisional
- Strategic
- Matrix
- Project

2.1 Functional Enterprise Structure

Functional structures within an enterprise have organization designs based on areas of specialization. This is basically the classical structure that has predominated throughout most of this century (Figure 6.1). Each element in the structure has a well defined activity and relationship with other organizational elements. According to Pearce and Robinson (Pearce and Robinson 1991, p. 329) there are advantages to a functional enterprise structure.

Table 6.1
Enterprise Functional Requirements

Organisation	R&D	Operations	Marketing	Administration
Religious Organization	Theological framework for a changing society	Conduction of services and provide support	Proselytizing	Manage collection and disbursement of funds, train new religious, educate, and manage facilities
Fast Food Franchise	Develop new recipes and management tools	Produce food for sale, maintain equipment and facilities, and work with suppliers	Advertise on TV and printed media, in-store promotions, and obtain new franchisees	Manage collection and disbursement of funds, recruit and train staff, and manage facilities
Academic Institution	Develop new courses, basic research to develop knowledge	Disseminate information through teaching and publications	Recruit students, seek research grants and contracts	Manage collection and disbursement of funds, recruit faculty and staff, and manage facilities
Government	Develop programs for good and welfare of society	Manage developed programs to provide necessary services to society	Keep citizens fully informed of all actions and make information readily available	Manage collection and disbursement of funds, recruit and train staff, manage internal and external relationships, manage facilities and resources
Technological Enterprise	Develop new technological innovations	Translate developed innovations into products, processes or services	Generate demand and provide interface with customers	Manage collection and disbursement of funds, recruit staff, manage facilities and resources

(Source: Modified from Heizer and Render 1991, p. 9.)

The advantages of a functional enterprise structure include efficiency through specialization, development of functional expertise, differentiation and delegation of decisions, central control of decision making, defined objectives, and ease of financial control.

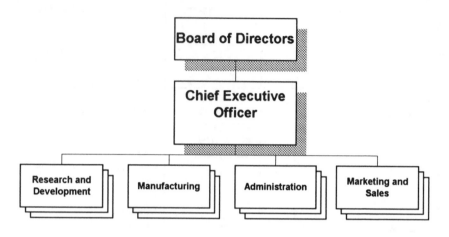

Fig. 6.1 Functional Enterprise Structure Example

Notwithstanding these advantages, this narrowly focused structure can be very limiting from an innovative viewpoint. Each element in the structure views the environment solely from the perspective of its structural character. Narrowness of specialization is a disadvantage in that a functional structure limits innovation. The structure promotes functional rivalry, adding to the disadvantage. Other disadvantages includes the difficulty in functional co-ordination and decision making, and potential for interfunctional conflict. This form of enterprise structure limits organizational learning due to specialization. Finally, functional organizational structure has a slow response to customer requirements since a customer's concerns may cover a number of functional areas.

Technological developments have forced many functional organizations to adapt to other enterprise structures which encourage technological innovation. As enterprises increase in size and technological complexity, emphasis has shifted to organizational design which facilitates the integration of multiple ongoing programs (Thamhain 1992, p. 44).

2.2   Geographic Enterprise Structure

A technological enterprise serving different geographic areas can use a geographic structural framework (Figure 6.2). Variation in the needs of these areas can frequently necessitate different approaches in implementing technology products, processes and services in a specific geographic environment (Pearce and Robinson 1991, p. 328). This is particularly obvious when the enterprise has organizational elements in different countries, as is increasingly the case in technology. A geographically structured enterprise easily responds to local market conditions.

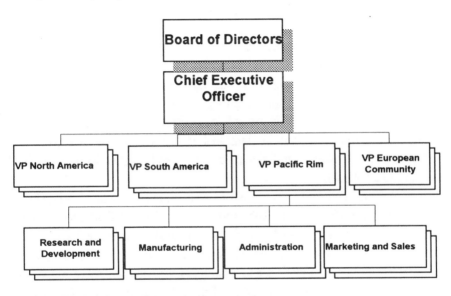

Fig. 6.2 Geographic Enterprise Structure Example

The advantages, according to Pearce and Robinson (Pearce and Robinson 1991, p. 330) of a geographically structured enterprise include closer customer relationships and responsiveness. Each geographical structural unit serves as a profit and loss center, thus allowing better control of decisions. Improved functional co-ordination in a geographic market provides another advantage. The use of a structure based on geographic areas provides for a local economic advantage in that the costs are based on the market served. This type of structure offers an enterprise with the ability for management training by rotation and thus provides for multi-cultural and geographic learning.

The disadvantages of geographic enterprise structure is the difficulty in communication and implementation of total enterprise objectives and goals. A geographically organized enterprise does not lead to organizational consistency, due to variations in existing cultural styles. The additional management layers and duplication of structures which arise to meet the diverse geographical nature of the enterprise can also be disadvantages. This problem can be reduced by relying upon modern communication structures such as the use of an enterprise Intranet[1] communication system. The geographic nature of the enterprise can lead to sub-optimization to meet local objectives and constraints. This sub-optimal reaction can cascade throughout the enterprise with possible significant chaotic results.

2.3   Divisional Enterprise Structure

This structure is associated with diversified products, processes or services which are provided by the enterprise for unrelated market channels, or heterogeneous customer groups (Figure 6.3). A top level functional structure usually cannot meet the increased co-ordination and decision making requirements resulting from a divisional structure (Pearce and Robinson 1991, p. 332). However, functional structure can be maintained for each of the divisional elements.

A divisional enterprise structure forces co-ordination to appropriate levels of the enterprise (Pearce and Robinson 1991, p. 331). Clear accountability for performance and co-ordination to appropriate levels are two specific advantages of a divisional enterprise structure. This type of structure has close relationships to suppliers and customers, providing a more responsive enterprise. The chief decision maker of the total enterprise can concentrate on overall enterprise management, with each division delegated responsibility for their element of the enterprise. Another advantage is that the divisional structure has a clear accountability for performance. The structure also retains advantages of functional specialization within each division. The divisional structure enables enterprise training through managerial rotation.

The disadvantages of this structure according to Pearce and Robinson (Pearce and Robinson 1991, p. 331) include the potential for dysfunctional resource competition. Delegation to divisional leadership teams can be

---

[1] An Intranet is a networked enterprise information system using the metaphor and interface of the Internet system, including connections to the Internet system.

uncertain and has a potential for policy inconsistencies. The divisional structure of an enterprise may make it difficult to accurately distribute overhead costs.

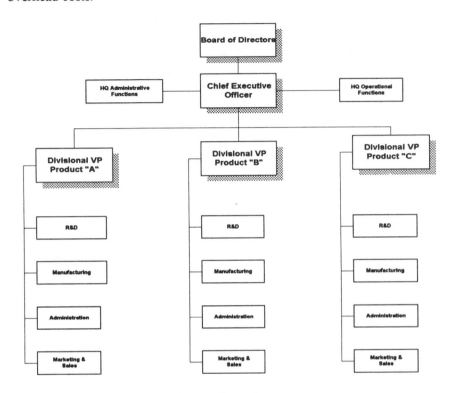

Fig. 6.3 Divisional Enterprise Structure Example

2.4   Strategic Enterprise Structure

As technology and environmental complexity increase, an enterprise may group elements according to strategic interests (Figure 6.4). This is done to improve strategy implementation, promote synergy and greater control (Pearce and Robinson 1991, p. 332). This is usually accomplished by the addition of another management layer with organizational units having similar strategic interests grouped under a strategic manager. These groups are called *strategic business units* and are based on independent product-market segments.

A strategic enterprise structure has advantages which include improved co-ordination between elements with similar strategic concerns and increased management control of large, diverse enterprises. This type of structure provides ease of distinct and in-depth enterprise planning with increased accountability. However, according to Pearce and Robinson (Pearce and Robinson 1991, p. 333), the disadvantages of a strategic enterprise structure include increased managerial complexity through the introduction of another management level with the potential for dysfunctional competition for enterprise resources. This structure can have difficulty in defining strategic managers' roles. The degree of autonomy delegated between strategic managers and divisional managers may cause conflicts within the enterprise.

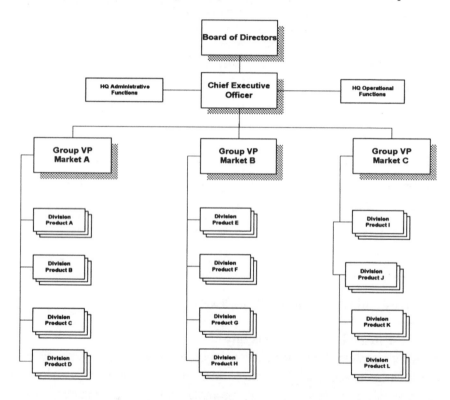

Fig. 6.4 Strategic Enterprise Structure Example

2.5 Matrix Enterprise Structure

Structural enterprise forms can be presented as a continuum ranging from functional to project structures (Archibald 1992, p. 44; Youker 1975). A matrix enterprise structure is used to describe project-driven two-dimensional enterprises and enterprises that have *permanent matrix* form (Figure 6.5). The objective of a matrix enterprise structure is to provide a means for a multifunctional approach to the technological development. In a matrix structure accountability, decisions and results are shared. The matrix enterprise provides for a dual channel of authority, performance responsibility, evaluation and control (Pearce and Robinson 1991, p. 333). The matrix structure is suitable for providing flexibility in human and physical resource allocation and forms a structure between a functional and a total project structure. The dynamics of technological change have led to the matrix structure for enterprises. The structural enterprise philosophy, in a matrix structure, reflects relationships rather than classical functional alignments.

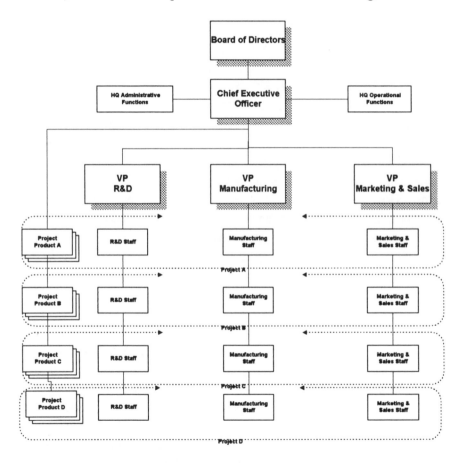

Fig. 6.5 Matrix Enterprise Structure Example

A matrix organization includes, according to Cleland and King (Cleland and King 1983, p. 279-281), four basic activities: functional support, project management, administration and strategic planning. *Functional support* is the activity for facilitating technology provided by elements from production, marketing, R&D and finance, while *project management* comprises activities to accomplish particular enterprise objectives or goals such as the development of a new technology product, process or service. The accommodating services related to mission activities are provided through the *administration* activity. *Strategic planning* provides the development of plans and strategies for new technology products, processes or services for the enterprise's future.

Matrix enterprise structure has both advantages and disadvantages according to Pearce and Robinson (Pearce and Robinson 1991, p. 334). The advantages of a matrix enterprise structure is that it includes a wide variety of enterprise activities and enables rapid response to enterprise environmental changes. The matrix enterprise structure maximizes human resource utilization. The structure fosters innovation, creativity, and enterprise learning. One of the disadvantages of matrix structures is that horizontal and co-ordination requirements can be excessive. A matrix structure can result in confused and contradictory policies.

Originally matrix structures did not assign R&D staff directly to matrix teams. However, with the advent of multifunctional teams it has become apparent that teams must contain representatives of all major organizational elements which interact in bringing a new technology innovation through all of its life cycle stages (see Chapter Five - *Technological Life Cycle and Decision Making*).

2.6 Project Enterprise Structure

An enterprise can be structured using projects as the basis for organizing. Project structures are common in large scale construction and governmental projects (Heizer and Render 1991, p. 692-696). In this structure, the units within the projects tend to be multifunctional and capable of operating as individual units with little or no support from the central enterprise. Large construction projects usually are formed for the design and construction of a building or complex. The United States federal government used this concept very successfully in the Manhattan Project which developed the first nuclear weapons. Similarly, NASA used this concept for the Apollo and Space Shuttle Projects.

Project management in this structure is focused on one basic objective and can lose focus of how this objective must nest within the overall enterprise objectives (see Chapter Two - *Technological Strategy*). An advantage of this type of organizational structure is the strong commitment from members to accomplish a major task. Fiscal control can more easily be maintained in this structure. In this framework, the central enterprise exercises overview but not direct control. This allows project managers to deal with achieving the overall project objectives once delegation has been made.

However, in this type of project structure, project problems may have major repercussions in the total enterprise. Project managers may come into conflict over policies with the enterprise leadership team, possibly causing loss of critical support. The project management structure can result in sub-optimization of enterprise resources. If the project does maintain both a project and enterprise focus, then the project goals may come into conflict with enterprise long term objectives.

3. ENTERPRISE COMMUNICATION

Communication is the cornerstone of technology enterprise success. Without the information obtained through clear channels of communication, the environmental *noise* introduced by both location and human elements can obscure the *true* information contained in any communication. There is a need to understand how enterprises *truly* communicate with both the internal and external environment under any structural model. It is also critical that the *sources* of enterprise noise factors are understood and mitigated.

The Safoutin and Thurston (Safoutin and Thurston 1993) model of communication for interdisciplinary design team management is one approach to structural communication analysis (Figure 6.6). The cost of technological flaws, in many instances caused by communication problems, increases the later that corrective action is taken in the stages of the life cycle (Figure 6.7). In a multifunctional or matrix team, communication occurs between members who represent different functional groups. The cognitive model of Safoutin and Thurston can be expanded beyond the design information process to cover the entire life cycle process.

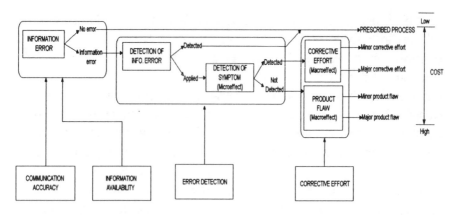

Fig. 6.6 Model of Interdisciplinary Communication (Source: Safoutin and Thurston 1993, © 1993 IEEE. Reprinted by permission.)

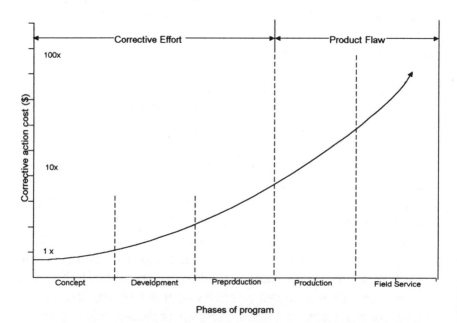

Fig. 6.7 Cost of Late Corrective Actions (Source: Safoutin and Thurston 1993, © 1993 IEEE. Reprinted by permission.)

The design information system includes external information which enters the system and includes various environmental data, some of which may have little relevance to the technological process (Safoutin and Thurston 1993). Internal information including requirements which are converted into descriptions of technical systems and information transmitted between various members of the enterprise structure and team members in a multifunctional context are elements of a design information system. Information processing activity undertaken by members of a multifunctional team and other enterprise elements are also within the design information system. A communication event is defined by the presence of an encoder (sender), a decoder (receiver), a channel (means of transmission), a referent (topic), and a message (Safoutin and Thurston 1993). The information process encounters barriers due to unavailability and inaccuracy of information communication. These barriers produce information errors by presenting incorrect or incomplete information. This can cause a cascading of incorrect and incomplete information throughout the enterprise, resulting in decisions which may be flawed or even catastrophic.

Safoutin and Thurston (Safoutin and Thurston 1993) proposed a number of strategies for addressing these barriers. One strategy is controlling information availability through *member centrality*, i.e., insuring closeness between team members. This strategy also encompasses *information centrality and initiation*, which Safoutin and Thurston define as using members who possess information prior to team formation.

Controlling communication accuracy, the next strategy, has a number of elements including *decentering and cognitive similarity*. This concept consists of making sure that members who communicate are at the same level of understanding about the technology they are communicating. *Rate*, i.e., the amount of information communicated, should not be excessive, and *channel availability*, or insuring familiarity and availability of the communication channel, e.g., scheduling system, Groupware, etc. are part of this strategic element. Communication accuracy is aided through:

- *Feedback* (response to communications)
- *Message complexity* (simple messages)
- *Organization* (simple structure)
- *Objectivity* (increased degree of objectivity of the communication, i.e., lack of importance of context)
- *Ambiguity* (unambiguous communications)
- *Referent complexity* (simple topic of communication)

The strategy for *controlling error detection* includes: insuring familiarity of members with assigned tasks (*familiarity*); *prominence of application*, i.e., the definition of critical tasks where externally communicated information is important; and *salience of symptoms*, or the definition of a critical task so that any external erroneous critical information will cause a small error. The final strategy of Safoutin and Thurston in removing the barriers to effective communications is through *controlling corrective effort*. This is achieved through detecting small errors before they can cause major problems and so that the impact occurs soon after the information is received (*immediacy of effect*).

Effective communication increases the likelihood of the success of technological development projects (Morelli et al. 1995). Communication effectiveness is dependent on enterprise structure and the type of technological project. There are various types of technical communications according to Morelli et al. These include: co-ordination, knowledge, and inspirational communication. *Co-ordination* is where technical information is transferred and used for task co-ordination. *Knowledge* communication includes consultative information, instruction and skill development. *Inspirational* communication is used for motivational and managerial affirmation objectives.

The majority of communication in technology product development was found by Morelli et al. to be co-ordinative (Morelli et al. 1995). Therefore, effectively selecting team members and creatively delineating organizational boundaries facilitate co-ordinative communication. High-frequency communication linkages were also observed by Morelli et al. Thus enterprise designs which facilitate these high-frequency linkages would improve the technological developmental process. Also a strong bi-directional upstream and downstream communication pattern assisted development of successful products.

While team communications are important and have been the subject of considerable research, successful enterprises have effective communications with all their employees (Young and Post 1994). Young and Post describe eight employee communication principles for successful enterprises. Communication must begin at the top of the leadership team, i.e., the CEO must be a champion of communication. Furthermore, the actions of the enterprise must match the words used in communication, thus reducing ambiguity in the communications process. The enterprise must have a true commitment to two-way communication. Communications at all levels of the enterprise should emphasize face-to-face communications. All levels of management should share in the responsibility for employee communication and not have it delegated solely to immediate supervisory management.

Enterprises should make sure that the bad news to good news ratio does not bias the communication process. Knowledge of the audience is all-important, i.e., knowing customers, clients and enterprise members. A major stakeholder group consists of the enterprise employees, and the enterprise should therefore have a definite strategy for communicating with these stakeholders. This strategy should include not only what is happening, but why and how it is happening. Communications with the employees must be timely and continuous. While most employees are concerned with immediacy, any communication should link the *big picture* with the *little picture*, i.e., expand the vision of the employees. Not all enterprise communications will be positive, so employees may have adverse reactions. The enterprise should not dictate the way people should feel about the communications, but must be prepared for the consequences of the communications.

Young and Post (Young and Post 1994) concluded that employee communication is a critical management process and not a set of products. Furthermore, effective employee communication practices should be consistent under all organizational conditions and every manager should be a communicator.

4.   ENTERPRISE CULTURES

Much has been written on how enterprises learn and transmit knowledge internally. The process of organizational learning, i.e., the process by which the enterprise becomes aware of the qualities, patterns and consequences of its own experiences and develops models to understand these experiences, can be differentiated into cultures (McGill and Slocum Jr. 1994). These cultures of enterprises are categorized as *knowing, understanding, thinking,* and *learning.*

A *knowing enterprise* is defined by McGill and Slocum as one in which there is one way to do things within an organization. A knowing enterprise changes incrementally in reaction to changes in its environment. This type of enterprise also exercises a high level of control, i.e., enforces conformity, has routine behaviors and is risk averse. Knowing enterprises are termed *learning disadvantaged.* IBM, of the 1950s through 1980s, was a knowing enterprise.

Increased foreign competition and customer dissatisfaction, combined with technological change, required enterprises to move from the knowing enterprise to a learning form where all enterprise members comprehend organizational values, and action should reflect this core understanding, i.e., an *understanding enterprise*. This form of enterprise wants its customers to experience the organization's cultural values in all interactions. Many U.S. utilities are this type of enterprise and can appreciate only those changes that are consistent with their ruling myths. However, the deregulation of the utility industry is likely to lead to a totally changed competitive environment which will likely cause these institutions to move from the understanding enterprise structure to a model which will allow for a more competitive posture.

In a *thinking enterprise* the overruling philosophy is "*if it's broken, fix it and fix it fast, but don't focus on why it's broken.*" Management practices are programmed, discrete and solution-oriented. This is exemplified by the inappropriate use of management concepts of total quality management ("TQM"), re-engineering, re-processing and other such quick-fix concepts; where a quick seminar leads to the major introduction of the concepts without an understanding of the total enterprise system implications. This is exemplified by the overuse of such principles as promoted by many organizational consultants and quickly adopted by many government agencies, e.g., overuse of *customer orientation* in some agencies within the federal government of the United States.

In a *learning enterprise*, the organization learns from its experiences and has an understanding how the organization has reacted to these experiences. The enterprise maximizes the learning that can be achieved from each of the organization's interactions with employees, customers, clients, vendors, suppliers and even competitors, and includes the way the information is collected, processed and used. In a learning enterprise, the primary responsibility of the leadership team is to foster learning. Employees are responsible for gathering, examining and using the information from the learning process.

Table 6.2 compares the various enterprise cultures. A learning enterprise has a clear and consistent learning culture characterized by:

- Openness to experience
- Encouragement of responsible risk-taking
- Willingness to acknowledge failures and learn from them.

Table 6.2
Comparison of Various Enterprise Structures

Element	Knowing	Understanding	Thinking	Learning
Philosophy	Dedication to the one best way: • Predictable • Controlled • Efficient	Dedication to strong cultural values which guide strategy and action. Belief in the "ruling myth."	A view of business as a series of problems. If it's broke, fix it fast.	Examining, enhancing, and improving every business experience, including how we experience.
Management Practices	Maintain control through rules and regulations, "by the book."	Clarify, communicate, reinforce the enterprise culture.	Identify and isolate problems, collect data, implement solutions.	Encourage experiments, facilitate examination, promote constructive dissent, model learning, acknowledge failures.
Employees	Follow the rules, don't ask why.	Use corporate values as guides to behavior.	Enthusiastically embrace and enact programmed solutions.	Gather and use information constructively dissent.
Customers	Must believe the enterprise knows best.	Believe enterprise values insure a positive experience.	Are considered a problem to be solved.	Are part of a teaching/learning relationship, with open, continuous dialogue.
Change	Incremental, must be a fine tuning to "best way."	Only within the "ruling myth."	Implemented through problem-solving programs, which are seen as panaceas.	Part of the continuous process of experience-examine-hypothesize-experiment-experience.

(Source: McGill and Slocum Jr. 1994, © 1994 IEEE. Reprinted by permission.)

Schein (Schein 1995) discusses how dialog can also assist in enterprise learning. Dialog is a central element of any enterprise moving toward

200

becoming a learning enterprise. Figure 6.8 shows the various ways members of an enterprise can communicate. Enterprise learning is not possible, according to Schein, unless some learning first takes place in the leadership team subculture. Dialog at the leadership team level is not sufficient for enterprise learning to occur. What is required is communication or dialog throughout the entire enterprise. Total enterprise dialog in some form is necessary and integral to enterprise learning.

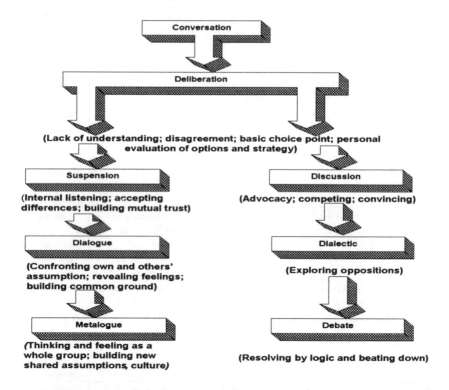

Fig. 6.8   Enterprise Communication (Source: Schein 1995, © 1995 IEEE. Reprinted by permission.)

4.1   Action Learning

An older concept than a *learning enterprise*, is an *action learning enterprise*, which was developed to deal with enterprises going through rapid change in their internal or external environment (Sankaran 1995). Action learning developed in England in the 1960s and is defined by Sankaran as:

> *"A process which brings people together to find solutions to problems, and, in doing so, develops*

*both the individuals and the organisation."
Action Learning is a potential tool for achieving
exchange of information and experience within a
rapidly changing environment, but is only a
portion of the activities required to change an
enterprise culture into a Learning Enterprise."*

The essential conditions required for successful action learning are regular meetings and support from the enterprise leadership team. The organization should, according to Sankaran, have a "Q" learning opportunity. "Q" learning is an opportunity to ask fresh and useful questions, when participants see not an idealized past but a challenging and possibly threatening future. Action learning usually begins by asking participants the following questions:

- What are we really trying to do here?
- What is preventing us from doing it?
- What can we contrive to do about it?

Locate the true stakeholders and ask the following questions:

- Who knows about what we are trying to do?
- Who cares about getting our solution implemented?
- Who has the power or owns the resources that can help us to implement the project?

Once this process is completed, an enterprise team can better understand the relationships and the factors which must be dealt with if a technological activity is to be successful under rapidly changing conditions.

4.2 International Enterprise Culture

Technological enterprises are, in many instances, being faced with operating in a multicultural international environment. This places new requirements upon an enterprise seeking to use a learning organizational archetype. These needs include an understanding of the cultural differences and how to effectively communicate and learn in this environment (Munter 1995). Figure 6.9 shows a mapping of the differences in high and low context cultures that managers of technology must understand in order to be able to operate effectively in an international environment. Cultures range from *high context*, i.e., establishing a context or relation first, to *low context*, i.e., getting right down to business. In an international technology environment, a manager must understand that various cultures have different attitudes on what determines communication credibility. This communication credibility is enhanced by the relative importance of the rank of the communicator. Relationships and personal track record or goodwill are of importance in the communication process. Many cultures place a higher value on expertise than on trust. Perceptions about image, i.e., age versus youth, male versus female, and class,are an important element in some cultures.The importance of shared values cannot be underestimated in communications. Language can also serve as a barrier to effective cross-cultural communication and learning. This barrier is caused, according to Munter, by:

- Semantics
- Word connotations
- Tone differences
- Perceptions

Barriers to effective multicultural enterprise interactions can, according to Munter, be overcome. It requires that managers of technology, who will be interacting with other cultures, develop an understanding of the culture associated with their enterprise. This can be achieved by reading about and discussing the culture before going to the country. While the manager is in the country, listening, reacting, and interpreting the culture is very important. It is important that technology managers encourage group members to learn by example, especially in relation to non-verbal communication, maintaining an open attitude of patience, tolerance, objectivity, empathy, and respect.

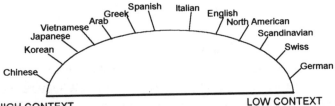

HIGH CONTEXT
* Establish social trust first
* Value personal relations and goodwill
* Agreement by general trust
* Negotiations slow and ritualistic

LOW CONTEXT
* Get down to business first
* Value expertise and performance
* Agreement by specific, legalistic contract
* Negotiations as efficient as possible

Fig. 6.9   Differences Between High and Low Context Cultures (Source: Munter 1995, © 1995 IEEE. Reprinted by permission.)

5.   ENTERPRISE LEADERSHIP

Top leadership support is a key factor in achieving technological development success (Green 1995). Figure 6.10, a model of top management support, shows the empirical relationships that Green determined from a study of two hundred and thirteen R&D projects in twenty-one major enterprises. The top leadership team is an active participant in individual R&D projects. An interrelated set of behaviors exists and is used by top management to support and *push* product and process development within the enterprise. However, involvement can lead to micromanagement of the project, an obvious negative factor. Top leadership team members are more likely to assume support roles when projects are directly tied to competitive issues.

One important question which this study raises is: does top management support exit for the right kinds of projects for the right reasons? The study also questions if top management should support new product or process research more than research aimed at incremental improvements. Furthermore, it questions if top management should support projects from the business side more than projects originating from R&D, especially when this support isn't related to the expected contribution of the project to the enterprise. Finally, are decisions driven by strategic concerns or do they represent biases? This implies that the top leadership team should carefully scrutinize the impact of their support and the manner of their support on projects. An example is John Scully's, the former CEO of Apple Computing, support of the Personal Digital Assistant ("PDA") concept, i.e., the *Newton*®.

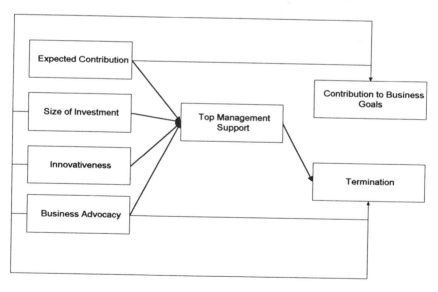

Fig. 6.10   Model of Top Management Support (Source: Green 1995, © 1995 IEEE. Reprinted by permission.)

6.   COMPETENCY

There are different types of competency which influence technological development success: management and total enterprise (Badawy 1995).

6.1   Management Competency

Managers of technology require various competencies depending on their level in the decision structure of the enterprise. The ingredients of managerial competency are given in Figure 6.11. Badawy shows in Figure 6.12 the managerial skill mix required at various management levels. According to Badawy (Badawy 1995), managers of technology are made, not born, and managing is a skill which can only be learned through practice. Badawy states that management is an applied social science and that there are no poor engineers or scientists, only poor managers. Therefore, the primary problems of managing technology are not technical, they are human.

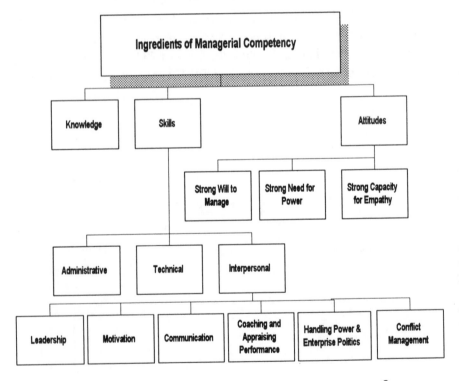

Fig. 6.11   Ingredients of Managerial Competency (Source: Badawy 1995, © 1995 IEEE. Reprinted by permission.)

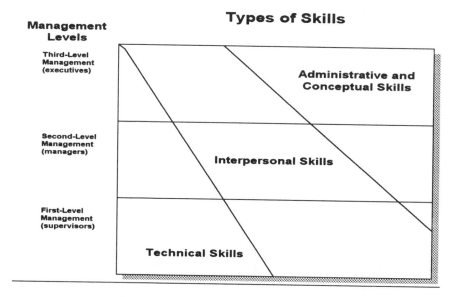

Fig. 6.12 Managerial Skill Mix (Source: Badawy 1995, © 1995 IEEE. Reprinted by permission.)

6.2   Enterprise Competency

Many enterprises have been structured around the concept of jobs, i.e., each individual is responsible for a specific job which involves specific accountabilities, responsibilities and activities; see, for example, the Dictionary of Occupational Titles (Lawler III 1994). Technological change and competitiveness issues are resulting in a trend for enterprises to become flatter through downsizing and reduced managerial layers. The impact of flattening an enterprise is to place increasing emphasis on the importance of individuals to self-manage and take performance responsibility. This results in the need to design enterprises in a manner in which the capabilities of individuals are the primary focus. The design should be such as to cause the enterprise human resources to be used in a manner that facilitates enterprise development so as to provide a competitive advantage. Consulting and other professional service firms are organized around the competency concept and are more flexible and more customer-focused than traditional job-based enterprises. Organizing around the competency concept requires both a focus on organizational core competencies and a focus on managing individuals and their capabilities.

7. INTERNAL FACTORS OF TECHNOLOGY

Technology development and applications can be impacted by internal and external factors. The external impacts are those caused by factors in the environment in which the technology is embedded (see Chapter One - *Technological Advancement and Competitive Advantage*). The internal impacts are caused by factors within the enterprise which the manager of technology may or may not have the ability to change. These internal factors can be grouped into:

- Financial
- Legal
- Organizational

Enterprises have control over the current technology by which they design, produce products and develop services. The manager of technology is often faced with change in large increments, i.e., *next-generation technology* (Edosomwan 1989, p. 38). These changes can have serious adverse impacts on an enterprise and place significant demands on a manager of technology. According to Edosomwan, management functions tend to move from the specific toward the general as technology changes and requires flexibility in managerial training and background. This movement from the specific toward the general may be one of the reasons why multifunctional teams have proved successful in developing new technology. As technology changes, the nature and activities of the manager of technology will change and evolve toward more participation through communication. One of the driving forces which requires greater flexibility of managers of technology is the rapidly shrinking technology life cycle. This cycle has been reduced from ten years in the 1960s to about eighteen to twenty-four months in the 1990s, and, it is estimated, to six months by the turn of the century.

In terms of shortened life cycle, in which the technology must be embedded, *high-velocity markets* (Slater 1994), encompass the embryonic and early growth stages of the technology life cycle. Technology markets are characterized by high growth and high investment requirements, unstable product technology, and changing competition. Hence, this requires managers of technology to deal with uncertainty and high risk (see Chapter Five - *Technological Life Cycles and Decision Making*). The greatest source of uncertainty is the lack of experience with the market and the new technology. Figure 6.13 shows the risk for all technological market participants.

Fig. 6.13 Risks for All Technological Market Participants (Source: Slater 1994, © 1994 IEEE. Reprinted by permission.)

The introduction of new technology by an enterprise is driven by opportunities and is constrained by existing technical knowledge. Numerous studies have indicated that an enterprise's risk of failure is greatly reduced when it operates in a familiar technological sector (Slater 1994). Moving too far from core competencies often results in increased development costs and lower or erratic entrant quality. An enterprise can position itself to be a *Technological Pioneer, Early Technology Follower,* or *Late Technological Follower.* Figure 6.14 shows the potential return of the three types of technological positioning.

208

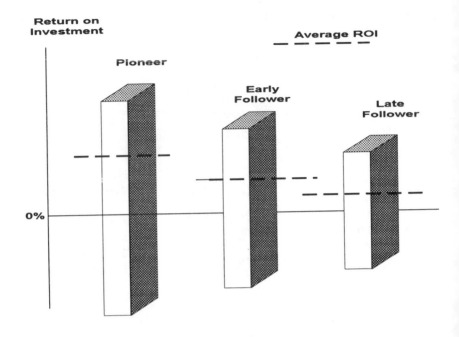

Fig. 6.14 Potential Return of Technological Positioning (Source: Slater 1994, © 1994 IEEE. Reprinted by permission.)

7.1 Technological Pioneer

Technological pioneer enterprises are innovators who are willing to take a higher risk for higher returns. This type of enterprise is the first to identify new market opportunities and has the technological capability to develop a new entrant that addresses these opportunities. The enterprise is willing to risk substantial resources to enter the market prior to its competitors. This type of enterprise reaps a competitive advantage from technological leadership, pre-emption of assets, and creating buyer switching costs. Pioneering enterprises have the opportunity to develop and position technology entrants for the largest and most lucrative market segments. Successful pioneering enterprises begin with a significant advantage over firms employing earlier technologies.

7.2 Early Technological Follower

An enterprise which positions itself as an early technological follower is a creative imitator (Slater 1994). The objective of this type of enterprise is to make incremental improvements that reduce costs and exploit unattended market opportunities. These enterprises have substantially reduced financial and market risk, which provides an attractive opportunity for growth and profitability. This substantial risk reduction results from lower costs than those incurred by pioneering enterprises. The early technological follower enterprise has an opportunity to assess how effectively the pioneering enterprise has positioned itself with respect to market needs and thus potentially to make appropriate adjustments. It should be noted that success for both pioneering and early technological follower enterprises requires rapid development and introduction of the new technological entrant.

8.   FINANCIAL FACTORS

All enterprises, private or public, require capital. Furthermore, the manager of technology must understand the utilization of capital resources. The manager of technology must be familiar with financial analysis, which includes the following factors:

- Capital
- Costs
- Pricing
- Budgeting

8.1   Capital Acquisition

Many established enterprises obtain capital additions through existing capital markets by loans or equity debt issues. It is the entrepreneurial enterprise which has limited choices. There are two primary sources of external capital for entrepreneurial enterprises, one visible and one invisible (Freer et al. 1994). The visible venture capital market in the United States is composed of over five hundred venture capital funds, which in 1993 were managing approximately thirty-five billion U.S. dollars in assets. The portion of capital from these professional venture funds, i.e., the visible venture market, is complementary to that from the private investors. These private investors comprise the invisible venture capital market which consists of a diverse and dispersed population of high net worth individuals interested in investing equity capital in entrepreneurial ventures. Financing from private investors is typically seed or start-up financing in the one to five hundred thousand U.S. dollars range raised from six to eight investors (Freer et al. 1994). The complementary relationship between the visible and invisible markets has two dimensions: stage and size. Stages of investment include:

- Start-up or seed capital
- Development or early-stage financing
- Pre-public offering financing

At all stages, venture capital funds tend to invest substantially more dollars than private investors. Size differences become larger as the stage of the financing advances. The private or invisible venture market preference is to invest at the start-up or seed stages while the venture capital funds tend to be more conservative. The private investors have a longer expectation horizon and are less risk-averse than the venture capital funds. At later stages, the size of the capital requirements can become much more than the individual investor is capable of supplying. However, to obtain the significantly higher amount of capital, i.e., between five hundred thousand and one million U.S. dollars, the enterprise must have demonstrated some success and moved beyond the higher-risk start-up stages.

Efficient markets allocate scare resources to the most productive uses; however, the invisible capital market is imperfect, due largely to the invisible nature of the informal venture capital markets. This means that seed and start-up capital is a scare resource traded in an inefficient market. Capital shortages can occur when an entrepreneurial enterprise attempts to obtain development financing for technology-based inventions. It is also difficult for enterprises to obtain financing for start-up and early-stage financing when the enterprise fails to meet size, stage and growth criteria. Equity financing when the enterprise is closely held and growing faster than internal cash flows can support is also difficult.

8.2 Cost Impacts

The rapidity of technological change resulting in shorter technological life cycles has serious cost impacts. These impacts can be both positive and negative. Accelerated product development ("APD") is a response to the shorter technological life cycles which can cause increased cost if not managed properly. APD consists, according to Crawford (Crawford 1993), of changing *strategic view*. This change in strategic view of the enterprise can assist in keeping costs impacts to a minimum. The change occurs when the enterprise seeks more incremental innovation and planned obsolescence, i.e., replacing products more frequently than demanded by the market, and results in dividing technology into engineer- and science-dominated development. The new APD strategic view must stress product quality at all points and reduce capital investments as much as possible. APD provides a means to have a quick response to changes in the marketplace.

New *management* methods are also used in APD. The use of *multifunctional teams* and *venture teams* with a strong leader and little formality and structure help in mitigating the problems associated with shorter technological life cycles. External acquisition will assist in meeting the challenger's due to rapid changes. New process technology such as computer-aided engineering ("CAE"), sterolithography, artificial intelligence ("AI"), integrated information systems, plus visual communication devices to increase quality communications, can assist in increasing the effectiveness of development.

New *administrative* support systems are part of APD. This can be given by instilling a corporate culture based on speed in everything without compromising quality. Another administrative approach is to use a lean and flat organization that stresses training and motivates workers with a reward system that includes equity positions if possible and means less structure in the operation of the enterprise. This new approach to administration requires new ways of *selling* R&D capability throughout the enterprise. The leadership team of the enterprise must establish firm and realistic deadlines.

However, APD can have a number of hidden costs (Crawford 1993). Opportunities for low-profit, trivial innovation that drives out the more profitable breakthrough types can also be the result of APD. This is a variation of Gresham's[2] financial law which states *"bad money drives out the good"* or, in terms of innovation, *"trivial innovation drives out breakthrough innovations."*[3] Mistakes which happen when skipping steps sacrifice necessary information and can cause increased cost. The major omission in the APD process is the acquisition of information. Human resource costs can increase when teams turn out to be inflexible. Unexpected inefficiencies resulting when the innovation process is under excess pressure will increase development cost in APD. Support requirements can in APD outstrip the enterprise's complex set of support resources, resulting in increased costs.

---

[2] **Gresham, Sir Thomas,** 1519-1579. An English financier and adviser to government of England. Sir Thomas has been traditionally but mistakenly credited with "Gresham's law," an observation that when two coins are equal in nominal value but unequal in intrinsic value, the one having the less intrinsic value tends to remain in circulation and the other tends to be hoarded.

[3] The author's restatement of "Gresham's law."

These hidden costs can be minimized through various actions (Crawford 1993). A clear and specific new technology strategy with management commitment and agreement can help reduce cost. The potential hidden costs can be mitigated by a constant review to determine if the development process is being successful. This review should include progress reports. A *lessons learned* approach to building learning on past APD projects will also reduce costs by possibly avoiding past errors. Similarly, a close customer interaction to determine if the APD project offers genuine benefits to the customer is an important element of the process. Independent assessment of APD project hidden costs will insure that costs are not being overlooked. Hidden costs can also be minimized by assessing existing system processes support and making necessary improvements.

The drive to accelerate technological developments and have a low cost position can be very difficult to obtain. Overall industry cost leadership, as a development strategy, can be in conflict with the APD project concept. According to Porter, a low cost position in an industry requires (Porter 1980, p.35-46) both organizational skill and resource requirements to be successful. The organizational requirements include having tight cost controls, frequent and detailed control reports, a structured organization and responsibilities, and incentives based on meeting strict quantitative targets. The skills and resource requirements include substantial capital investment and access to capital process engineering skills, intense supervision of labor, products designed for ease in manufacture, and a low-cost distribution system. Satisfaction of these requirements results in a particular leadership style, corporate culture and environment which can cause stress in an APD project system.

An important element in cost is the need to utilize new technologies and processes to assist in the overall developmental process. Many CEOs face a decision on whether to expend capital for ancillary new technology to assist the developmental process within their enterprises or to rely on existing technologies. Traditional financial methods for investment evaluation fall short in measuring the real merit of investing in a new technology and in evaluating the strategic economics of new technologies (Porter 1980, p.35-46). The declining prices of new technologies often lead decision makers, such as CEOs or leadership teams, to postpone capital investments and wait for lower prices (Eden and Ronen 1993). Figure 6.15 shows the declining price paradox facing these decision makers. There are several complications in this declining-price paradox by Eden and Ronen. First, the learning time should always be taken into consideration whenever examining an investment in a new technology. Secondly, acquiring new technology may provide an advantage whenever an enterprise goes on to a new generation of technology by reducing learning time and enabling earlier use of the new technology. Finally, new technology price reduction is sometimes a mirage in terms of both tactical and strategic considerations.

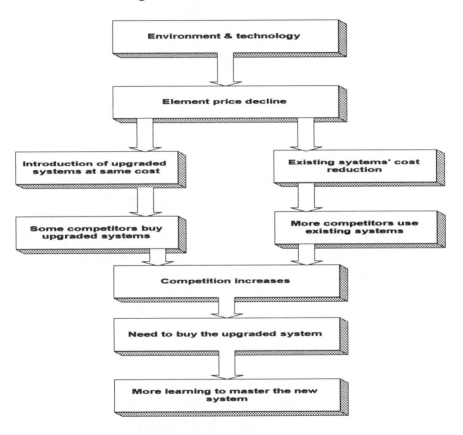

Fig. 6.15 Declining Price Paradox Facing Decision Makers (Source: Eden and Ronen 1993, © 1993 IEEE. Reprinted by permission.)

8.3 Pricing

The pricing of a new technology product, process or service is often a major enterprise problem. The managers of technology sometimes feel that pricing is outside their control (Dolan 1995). Determining the correct price can have a significant impact on profitability, e.g., an enterprise with an eight percent profit margin can increase this margin to twelve percent by raising the price by one percentage point. A better determination of the appropriate price for a new technology market entrant can be achieved by assessing value to the customer of a new entrant by careful market research and obtaining information through employee-customer interactions. It is important to develop an understanding of how customers determine value so that price customization can be performed. This value variation and opportunity for price customization can be established by determining, according to Dolan (Dolan 1995), the intensity of use, i.e., heavy users usually value a product more than light users. The price can be developed by determining use differentiation since customers have different uses for the same product, e.g., computers for home users versus office applications, and thus as a consequence different perceived values. Differences in performance value between customers can also help develop pricing, since those customers seeking higher performance place a higher value on that performance, e.g., notebook market differences due to types of screens, processors, etc.

Determining customer price sensitivity is another technique for developing pricing. The factors affecting customer price sensitivity include customer economics, customer search and usage, and competition. Customer economics depend on whether the decision maker is acquiring the technology for himself/herself or others, the cost relationship to total expenditure, and cost versus quality factors. Customer search and usage are related to expenditure of time for acquisition of the technology and relationships to the supplier. Questions related to competitive offerings and intangibles will impact price sensitivity. Acquirers of technology may be willing to spend more if additional incentives such as longer warranty periods are included.

Identification of an optimal pricing structure is an important enterprise activity. This includes creating a pricing structure to offer quantity discounts and when approving bundled[4] pricing. Pricing decisions must consider the reaction of competitors to potential pricing alternatives. Price wars can result from poorly designed pricing actions, causing reduced profit margins for all the participants. The pricing decision must also consider second and third order effects. Technology enterprises must monitor prices realized at the transaction level. The total set of pricing terms and conditions an enterprise

---

[4] Bundling refers to the practice of including other products and services.

offers can be quite elaborate. Enterprise technology managers should analyze the full impact of the pricing strategy, measuring and assessing the bottom-line impact. Pricing considerations include assessing how customers respond to technology prices, including the long-term effects of the customers' emotional reaction as well as the short term economic outcome. Some customers require more service than others; this requires analyzing revenue versus cost to serve the customer. High cost-to-serve a particular set of customers can have a very positive impact if the price these customers pay is high enough, e.g., value-added programming.

8.4 Budgeting

If a new technology project is to be initiated, current and future returns, adjusted to reflect the time value of money, must exceed investment. The decision to proceed with a new technological development for an undeveloped market requires an estimation (Howard Jr. and Guile 1992, p. 43-44) of the costs of developing a prototype and manufacturing a product for which neither a prototype nor a manufacturing line exists. An estimate of the market for, and sales of, a product, process or service that does not yet exist over the next several years should be made by the technology manager. Other uncertainties about such things as the long-term environmental effects of manufacture (ISO 14000), and the disposal and the potential liability exposure of making and selling the new entrant must also be considered.

Budgetary decisions should consider these factors when allocating scare enterprise resources. Techniques in designing and estimating a project concept within a budgetary constraint include the design-to-cost approach (Thamhain 1992, p. 154). Using this approach, cost targets are established for all major elements and then re-negotiated. The steps, according to Thamhain, in this approach include:

- Define total project target cost.
- Develop a work breakdown structure cost model.
- Establish target budgets for labor, material, travel, etc.
- Allocate related percentage efforts to all project subsystems.
- Establish a worker-hour target budget for each subsystem based on the percentage allocation of the total budget.
- Distribute all target cost information to all the team members.
- Ask functional managers to estimate the work force needs and other costs together with their task teams.

- Estimators provide:
  - best estimate of efforts
  - cost drivers
  - cost-performance trade-off analysis
  - alternative baselines for given target budget
- Analyze cost-performance trade-offs.
- Re-compute the related efforts for all subsystems and trade-offs.
- Select cost-driver subsystems of the project and work-out alternative solutions.
- Re-negotiate until a satisfactory solution to all elements is achieved.

8.5   Financial Analysis

A manager of technology should be versed in financial analysis, which is an important tool in assessing the strength of an organization. This tool provides the manager of technology with a measurement of how the enterprise is doing in comparison with its performance in the past and with the performance of competitors in the industry. Financial ratios (see Chapter Five - *Technological Life Cycles and Decision Making*) are an excellent analytical tool for analysis of the enterprise. Figure 6.16 shows these various ratios. These financial ratios include:

- Liquidity - Indicates ability to meet short-term obligations.
- Leverage - Indicates source of capital, i.e. owners versus creditors.
- Activity - Indicates how effectively the enterprise is using its resources.
- Profitability - Indicates how effectively the total enterprise is being managed.

A manager of technology should have familiarity with the following financial terms (Thamhain 1992, p. 232-234):

- Activity Accounting - Recording data by a specific organizational segment.
- Activity Base - Measure of an operating activity; this is used to allocate indirect cost.
- Budget or Cost Variance - Difference between actual costs and project costs.

- Capital Budgeting - Process of long-range planning involved with major additions or reductions.
- Cost Centers - Clustering of costs by functional areas.
- Cost Flow - Accounting method to classify process data to show product and period costs.
- Labor Rate Variance - Measure of the ability to control wage rates and labor mix, i.e., the difference between actual wage rate and standard wage rate multiplied by the actual hours worked.
- Lead Time - Interval between the time an order is placed and the time of arrival and use.
- Learning Curve - Mathematical expression of the fact that labor time will decrease at a constant percentage over doubled output quantities.
- Line-time budget - Accountability for expenditures is identified with specific expenditure lines in the budget.
- Overhead Rate - Allocating indirect costs to products, creating an average overhead cost per unit of production activity.
- Period Costs - Costs not inventoried and treated as an expense in the period in which incurred.
- Performance Budget - Adjusted budget prepared after operations to compare actual results with revenues and costs that should have been incurred at the actual level attained.
- Variance Analysis - Investigation of the causes of variances in a standard costing system.

8.6   Financial Appraisal Techniques

The financial appraisal techniques that are useful to a manager of technology include *Net Present Value ("NPV"), Internal Rate of Return ("IRR"),* and *Payback Period.* The NPV is the sum of all future cash flows discounted at the required rate of return minus the present value of the cost of investment. The IRR is the rate of return which equates the anticipated net cash flow with the initial outlay. A project can be acceptable if its IRR is greater than the required rate of return and the payback period is the length of time required to repay the initial investment. These are some of the major financial appraisal techniques and not the total set of these techniques. It is important that managers of technology apply as many of the available techniques as appropriate to manage new technology projects.

218

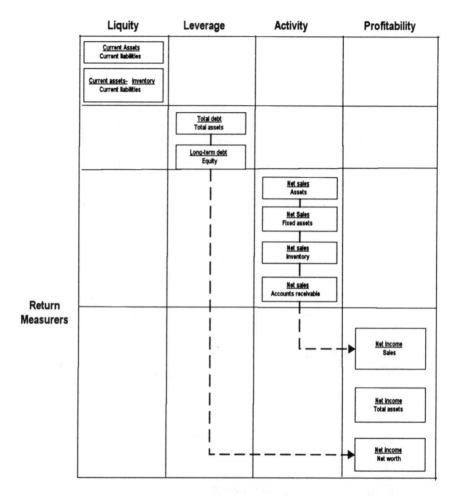

Fig. 6.16 Price Rations (Source: Pearce and Robinson 1994, p. 198. Reprinted by permission.)

8.7   Hurdle Rates

Hurdle rates are those rates of return above which a decision maker is willing to risk any new investment. Most enterprises use hurdle rates to determine if an investment into a new R&D project, facility or other investment is acceptable when all risks are considered. These hurdle rates are also related to the *time horizons* of the enterprise in terms of acceptable time to recapture investments. The decisions on hurdle rates and time horizons have a serious impact on development decisions for new technology projects. The question of a time horizon of U.S. enterprises has been an important issue throughout discussions on competitiveness with foreign firms (Poterba and Summers 1995).

Poterba and Summers found that enterprises in the United States employ hurdle rates in their capital budgeting procedures that are higher than standard cost-of-capital. Average discount rates applied to constant cash flows were 12.2%. Another result was that the U.S. enterprises and their decision makers had systematically shorter time-horizons than their European and Asian competitors. It appears that some governmental policy factors exert a powerful effect on corporate planning horizons. These factors are corporate tax rate, R&D tax credit, corporate tax deduction for dividend payments and a stable tax policy. The factors which do not seem to impact planning horizons include investment tax credit for equipment, capital gains tax reduction, corporate control regulations, and tax on short-term trading of corporate equity issues.

9.   LEGAL FACTORS

Generally, new technological developments will be influenced by one or more laws or regulations with which the new technology product, process or service must comply. If the new technology entrant is to be employed in other countries, the laws and the regulations of those foreign nations may also impact the project. Laws and regulations may impact the acceptability, entrant operation, development processes or any combination of these factors. The manager of technology has to consider the legal constraints arising from laws and regulations governing environment, intellectual property and patent (see Chapter Seven - *Technology Transfer*), contract, employment, corporate, taxes, financial accounting, securities, exports and imports including the Uruguay Round and the World Trade Organization ("WTO"), and government procurement. Each of these can form significant barriers to the implementation and successful introduction and growth of new technology.

It is important that potential laws and regulations be considered as early in the project as possible since they can have significant impact. It is possible that existing laws and regulations may offer a barrier which cannot be overcome, thus curtailing a particular development vector. Also, if legal barriers are recognized early in the process, it may be possible to develop alternatives. This would require actively seeking acceptable solution space for the development process before resources are expended. One technique is structured constraint deconfliction, a systems engineering technique, which is an organized way to challenge an overconstrained condition (Grady 1995, p. 220).

9.1 Enterprise Legal Structure

A technology enterprise can have numerous legal structures. These structures include:

- Partnership
- Limited Liability Company
- Corporation
- Consortium
- Joint Venture

The choice of structure depends on the objective of the new enterprise. Enterprises, which arise from entrepreneurial push, usually take forms which will entice investment. The choice of form depends upon the enterprise objectives and the prevailing tax laws. If the tax laws allow the investors to recoup losses easily then one form, such as a R&D partnership or limited liability company, would be employed. If the objective of the new enterprise is not just to develop a particular technology and obtain the economic rents by selling off or licensing the technology, then a more standard corporate form would possibly be utilized.

In many instances, groups of corporations will join for a particular technological development. They may form a consortium where individual members maintain their individuality, such as was used in the development of the PowerPC®. There are some cases where the participants form a joint venture and also have equity positions. These equity positions would be vested in other corporate enterprises. Each type of enterprise legal structure has advantages and disadvantages.

A limited partnership consists of at least one general partner and a number of limited partners. The general partners carry all the risk and liability equality. The limited partners only have their investment at risk. Partnerships are taxed so that each participant is taxed on ownership percentage. However, loss pass-through is severely limited by existing tax laws in the United States.

Limited liability companies are organized similarly to standard corporations but treated as a partnership for federal tax purposes, i.e., individual equity participants are responsible for their portion of the corporate tax. Income and capital gains both from operations, licensing or the sale of technology, and gains from disposition of the enterprise, will not be taxed at the enterprise level. Member-managers must hold at least one percent of the capital and profit interest and must meet certain capital requirements. Losses from R&D and other corporate expenses are subject to the at-risk and passive activity loss tax limitations in the United States. All members are protected to the extent of their equity investment. Members can contribute cash, property, promissory notes and services rendered for a profit interest in the enterprise without realizing taxable income. Personal holding company rules which are applicable to closely held corporations are avoided. Long-term capital gains in the treatment of the sale of patents created by personal efforts of individuals in the enterprise are allowed by U.S. tax laws. By U.S. law, venture capital investors and valuable employees and suppliers with an interest in the venture tied to success or to a particular project are allowed.

There are two types of standard corporation, i.e., "C" or standard corporation and "S" or corporation organized under subchapter S of uniform U.S. corporate law. The "C" corporation is a standard form of a corporation with equity owners. The corporation is taxed at the enterprise level, resulting in a double taxation of the distribution of profits to the shareholders. Losses in "C" corporations may not be passed through to the equity participants, i.e. the shareholders by U.S. tax code. "S" or Sub-chapter corporations are generally not taxed at the enterprise level in the United States. "S" corporations must adhere to restrictions not applicable to limited liability companies. These restrictions on "S" corporations include that acquisition of shares by non-qualified shareholders and the existence of more than one class of stock. Non-qualified shareholders include domestic and foreign corporations, trusts and foreign individuals. Under U.S. tax code, losses by "S" corporations may be passed through to shareholders, but are limited to the amount of capital contribution and loans.

10.   ORGANIZATIONAL FACTORS

Organizational structures reflect patterns of internal relationships and, when combined with internal processes, determine the internal *technology delivery system* (Porter et al. 1991, p. 308). Technology and the changes associated with that technology can make large demands on an enterprise. How the organization is impacted and responds to these changes is critical for the management of technology.

An organization has a set of internal understandings of how human and other internal resources are utilized to achieve the objectives of the enterprise (Gilman 1969). Organizations have a structure of understandings ranging from policies and procedures to members' assignments. The enterprise is an active agency where processes and structure are related to the efforts of members in the joint accomplishment of objectives. Thus an enterprise consists of, according to Gillian (Gilman 1969):

- Explicit or implicit objectives
- Formal and informal patterns of authority and responsibility
- Human and non-human resources
- Constant interaction of subsystems

The study of organizations has been developing from the early principles of Taylor[5] through the current concepts of re-engineering and complexity. Taylor and his associates emphasized making internal operations as efficient, rational and predictable as possible. Henri Fayol[6] identified a set of organizational principles more encompassing than Taylor's which served as guidelines to design an appropriate structure. These principles included:

- Division of labor
- Unity of command
- Unity of direction
- Subordination of individual interests to the common organizational goal.

Henri Fayol was one of the first management scientists to use the term *departmentation*, which relates to segregating enterprise resources to departments based on purposes, processes, customers, geographic area, and other factors. Some of the more recent organizational factors which have been considered during the 1950s and early 1960s are: centralization, decentralization, functional, departmental, product, process, geographical, and line and staff. In the late 1960s and 1970s the organizational factors

---

[5] **Taylor, Frederick Winslow,** (1856-1915), an American industrial engineer who is considered the originator of scientific management in business. Taylor's management methods were published in *The Principles of Scientific Management* (1911), and his name became synonymous with the efficiency movement.

[6] **Fayol, Henri,** a French management scientist published *Industrial and General Administration* (1925) and *Industrial and General Management* (1949).

included: product and project management, system analysis, and technological forecasting. In the 1980s the organizational activities shifted to the concepts of Total Quality Management ("TQM"), innovation, self-assessment, and matrix organization. Beginning in the 1990s, the principal focus has been re-engineering, and multifunctional teams. Each decade seems to bring new management concepts, each searching to improving the function of the enterprise. These concepts have been associated with various management and organizational tools. Figure 6.17 shows these various tools and the percentage of enterprises using them and the degree of satisfaction with the results of these tools (Rugby 1995).

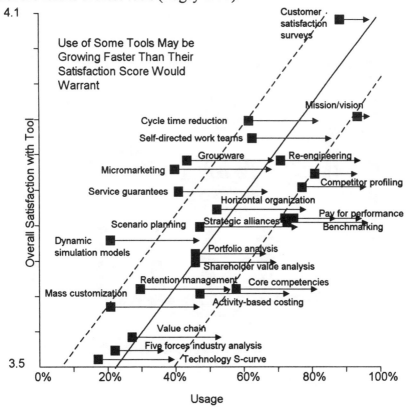

Fig. 6.17 Various Management Tools and Degree of Satisfaction (Source: Rugby 1995, © 1995 IEEE. Reprinted by permission)

The structural analysis has formed a basis for annualizing organizational factors. According to Mintzberg and Quinn (Mintzberg and Quinn 1991, p. 331-332), an organization has six basic components: strategic apex, operating core, middle line, technostructure, support staff, and ideology. *Strategic apex* is where the whole enterprise is overseen and the chief decision maker or leadership team resides. The *operating core* is the base of the enterprise where both human and non-human resources reside and perform the basic enterprise work. The hierarchy of authority between the operating core and the strategic apex is the *middle line* of the organization. The *technostructure* consists of analysts outside the hierarchy of line authority; similarly, the *support staff* provide various internal services, but are outside the line of authority of the organization. The binding force, according to Mintzberg and Quinn is the *ideology*, i.e., strong *culture* encompassing the traditions and organizational beliefs. Figure 6.18 shows a diagram of this structure.

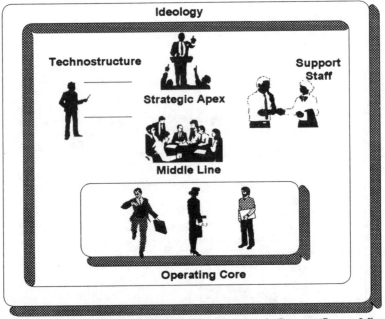

Fig. 6.18 Organizational Model Based Upon Mintzberg's Concept (Source: Mintzberg and Quinn 1991, p. 332)

Enterprises have various activities associated with their technological core (operating core), administrative actions, and boundary-spanning needs (Sebora et al. 1994). There are three levels of human resources in organizations: strategic level (Strategic Apex), middle management (Middle Line), and operational level (Operating Core). Successful technological innovation requires each organizational level to be involved in different, but interdependent, ways. In entrepreneurial enterprises, the strategic level takes the greatest responsibility for the innovative activity of their enterprises. The top technology managers are responsible for initiating and controlling change within their enterprises. According to Sebora et al. (Sebora et al. 1994), these members of the leadership team, who are intimately involved in both the market-related (external) and the product technology-related (internal) activities, believe that they are more innovative than enterprise members at other levels, even if they are not.

Middle line or middle managers may be able to generate technological and management innovations due to the synergy in their role serving as the link between the strategic apex and the operational levels in the enterprise. These technology managers also have the ability to facilitate innovation due to their bridging role in communication and resource allocation. The operational level members also have a role to play in technological innovation. These members not only serve to implement innovations but can also generate technological innovations. To accomplish this requires empowering operational level members to innovate. In many enterprises, operational core members are generally considered implementers of innovations, rather than the contributors to the enterprise's innovation process (Sebora et al. 1994).

The structure of an enterprise can facilitate innovation by providing a structure which encourages innovation (Bart 1995). The structure of an enterprise is concerned with the enterprise design and defines the roles of its members as well as their reporting, responsibility, and accountability relationships. Formal systems and procedures are the hallmarks of large enterprises. One way for improving innovation and intrapreneurship[7] in large organizations is for them to become less formal in their operations and systems, and to reduce the number of rules, policies, procedures, and controls Table 6.3 shows the difference in rules, policies, procedures and controls in managing new and established operations.

10.1 Organizational Activity Analysis
Organizational activity analysis approach provides managers of technology with a technique for diagnosing an organization's functioning, in terms of justifying costs and demonstrating value (Leifer and Burke 1994).

---

[7] Intrapreneurship is the exercise of entrepreneurial qualities within a formal existing enterprise.

Organizational activity analysis examines unit mission-related functioning through the use of an activity database. This form of analysis can be utilized to deploy members consistent with priorities and reward members consistent with work performed regardless of function. Activity analysis can insure all activities are consistent with the enterprise's mission and reduce redundant, fragmented or misplaced work. The organizational activity process improves direct-to-indirect cost ratios. The process also simplifies and rationalizes the enterprise structures. The data collected during the organizational activity process must be updated and maintained if the process is to be effective.

Table 6.3
Managing New and Established Operations

Variables	Product Type	
	New Product	**Established Product**
Rules and procedures	Few	Many
Adherence to rules	Low	High
Subordinate formal job definition	Broad/Low	Narrow/High
Budget tightness	Low	High
Plan/budget detail	Low	High
Written instructions	Few	Many
Reporting frequency	Low	High
Personal contact	Low	High
Customer contact	Low	High
Subordinate autonomy	High	Low
Intolerance for failure	Low	High
Amount of attention	Low	High
Reliance on other formal systems	Low	High
Reliance on other informal systems	Low	High
Overall tightness of control	Low	High

(Source: Bart 1995, © 1995 IEEE. Reprinted by permission.)

10.2 Technology Induced Organizational Stress

The pressure to innovate and transform can induce organizational stress. Technological change often translates into demands on enterprise members at all levels, resulting in dysfunctional outcomes termed *technology-induced stress* (Lee et al. 1995). The first stage of stress is a subconscious uneasiness referred to as *tension*. The next state is a conscious uneasiness referred to as *anxiety*, i.e., an awareness of tension experienced. The final state results from a specific known and immediate threat and is known as *fear*. Some of the results of technology-induced stress , according to Lee et al., include:

- Boredom
- Depression
- Anxiety
- Mental strain
- Job pressure
- Lower job satisfaction
- Lower quality relationships
- Absenteeism
- Physiological problems

The outcome of technology-induced stress impacts an organization's ability to rapidly respond to changing environments. It is important that enterprises develop methods for reducing technology-induced stress and yet maintain the response required to be successful in a rapidly developing technology market. One method of dealing with technology-induced stress is through the use of organizational demography. Research suggests that an organization's demography alters interaction and communication among organizational members (Lee et al. 1995). It is possible to change the dysfunctional consequences of a certain demographic distribution by changing the demographic mix among the following factors: age, sex, education, and tenure. While technology-induced stress can be related to these factors, laws and regulations within the United States related to the same factors may limit an enterprise's ability to make adjustments which adversely impact certain enterprise members. Nevertheless, it is important that an enterprise develop methods for dealing with technology-induced stress.

REFERENCES

Archibald, R. D. (1992). *Managing High-Technology Programs and Projects*, John Wiley & Sons, Inc., New York, NY.

Badawy, M. (1995). "Preventing Managerial Failure Among Engineers and Scientists." *IEEE Engineering Management Review*, 23(Summer), 58 - 65.

Bart, C. K. (1995). "Gagging on Chaos." *IEEE Engineering Management Review*, 23(Summer), 41 - 49.

Blanchard, B. S., and Fabrycky, W. J. (1990). *Systems Engineering and Analysis*, Prentice-Hall, Englewood Cliffs, NJ.

Cleland, D. I., and King, W. R. (1983). *Systems Analysis and Project Management*, McGraw-Hill, New York, NY.

Crawford, C. M. (1993). "The Hidden Costs of Accelerated Product Development." *IEEE Engineering Management Review*, 21(Summer), 21 - 28.

Dolan, R. J. (1995). "How Do You Know When the Price is Right." *Harvard Business Review*(September-October), 174 - 183.

Eden, Y., and Ronen, B. (1993). "The Declining-Price Paradox of New Technologies." *IEEE Engineering Management Review*, 21(Winter), 34 - 39.

Edosomwan, J. A. (1989). *Integrating Innovation and Technology Management*, John Wiley & Sons, New York, NY.

Freer, J., Sohl, J. E., and Wetzel, W. E. (1994). "The Private Investor Market for Venture Capital." *IEEE Engineering Management Review*, 22(Fall), 91 - 97.

Gilman, G. (1969). "The Manager and the Systems Concept." *Business Horizons*(August), 19-28.

Grady, J. O. (1995). *System Engineering Planning and Enterprise Identity*, CRC Press, Boca Raton, FL.

Green, S. G. (1995). "Top Management Support of R&D Projects: A Strategic Leadership Perspective." *IEEE Transactions on Engineering Management*, 42(August), 223 - 232.

Heizer, J., and Render, B. (1991). *Production and Operations Management: Strategies and Tactics*, Allyn and Bacon, Boston, MA.

Howard Jr., W. G., and Guile, B. R. (1992). *Profiting from Innovation: The Report of the Three-Year Study from the National Academy of Engineering*, The Free Press, New York, NY.

Lawler III, E. E. (1994). "From Job-Based to Competency-Based Organizations." *IEEE Engineering Management Review*, 22(Fall), 83 - 90.

Lee, T. S., Foo, C. T., and Cunningham, B. "Role of Organizational Demographics in Managing Technology-Induced Stress." *IEEE Annual International Engineering Management Conference*, Singapore, 38 - 43.

Leifer, R., and Burke, W. J. (1994). "Organizational Activity Analysis: A Methodology for Analyzing and Improving Technical Organizations." *IEEE Transactions on Engineering Management*, 41(August), 234 - 244.

McGill, M. E., and Slocum Jr., J. W. (1994). "Unlearning the Organization." *IEEE Engineering Management Review*, 22(Summer), 36 - 43.

Mintzberg, H., and Quinn, J. B. (1991). *The Strategy Process: Concepts, Contexts, Cases*, Prentice-Hall, Inc., Englewood Cliffs, NJ.

Morelli, M. D., Eppinger, S. D., and Gulati, R. K. (1995). "Predicting Technical Communication in Product Development Organizations." *IEEE Transactions on Engineering Management*, 42(August), 215 - 222.

Munter, M. (1995). "Cross-Cultural Communication for Managers." *IEEE Engineering Management Review*, 23(Spring), 60 - 68.

Pearce, J. A., II , and Robinson , R. B., Jr. (1991). *Strategic Management: Formulation. Implementation, and Control*, Richard D. Irwin, Inc., Homewood, IL.

Porter, A. L., Roper, A. T., Mason, T. W., Rossini, F. A., and Banks, J. (1991). *Forecasting and Management of Technology*, John Wiley & Sons, New York, NY.

Porter, M. E. (1980). *Competitive Strategy: Techniques for Analyzing Industries and Competitors*, The Free Press, New York, NY.

Poterba, J. M., and Summers, L. H. (1995). "A CEO Survey of U.S. Companies Time Horizons and Hurdle Rates." *Sloan Management Review*(Fall), 43 - 53.

Rugby, D. K. (1995). "Managing the Management Tools." *IEEE Engineering Management Review*, 23(Spring), 88 - 92.

Safoutin, M. J., and Thurston, D. L. (1993). "A Communications-Based Technique for Interdisciplinary Design Team Management." *IEEE Transactions on Engineering Management*, 40(November), 360 - 372.

Sankaran, S. "Introduction of Action Learning to Develop Engineering Managers in a Business Enterprise in Singapore." *IEEE Annual International Engineering Management Conference*, Singapore, 26 - 30.

Schein, E. H. (1995). "On Dialogue Culture, and Organizational Learning." *IEEE Engineering Management Review*, 23(Spring), 23 - 29.

Sebora, T. C., Hartman, E. A., and Tower, C. B. (1994). "Innovative activity in small business: Competitive context and organisation level." *Journal of Engineering and Technology Management*, 11(December), 253 - 272.

Slater, S. F. (1994). "Competing in High-Velocity Markets." *IEEE Engineering Management Review*, 22(Summer), 24 - 29.

Thamhain, H. J. (1992). *Engineering Management: Managing Effectively in Technology-Based Organizations*, John Wiley & Sons, New York, NY.

Youker, R. (1975). "Organizational Alternatives for Project Management." *Project Management Quarterly*, 8(March).

Young, M., and Post, J. E. (1994). "Managing to Communicate, Communicating to Manage: How Leading Companies Communicate with Employees." *IEEE Engineering Management Review*, 22(Spring), 24 - 31.

DISCUSSION QUESTIONS

1. How would you improve co-ordination in a government agency based on the concepts presented in this chapter.

2. How would you motivate workers in a high-technology environment? Is this different for a small entrepreneurial enterprise versus a government laboratory? - Explain.

3. Keep track of your organizational communication for one day using the following reporting formats. Circle the communication activity for each communication event

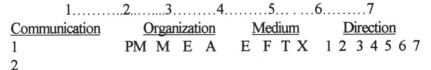

```
          1..........2........3.........4.........5......6.........7
Communication     Organization     Medium      Direction
1                 PM  M  E  A    E  F  T  X   1 2 3 4 5 6 7
2
.
```

where:

<u>Organizational Legend</u>

PM = Project Manager   M = Manager   E = Engineer   A = Administrative

Survey of Communication

<u>Medium Legend</u>

E = Email   F = face to face   T = telephone   X = FAX

<u>Direction Legend</u>

1 = Information required by yourself

4 = Information jointly required

7 = Information required by other person

4. Comment upon how a manager of technology in the following enterprises would handle financial factors:

- Entrepreneurial enterprise
- Government laboratory
- Government agency
- Multinational enterprise

8. Identify and discuss entrepreneurial qualities that are sources of potential legal liabilities in managing technology.

9. Comment upon the difficulties involved in forming teams within an enterprise from various functional areas. How would you go about selecting a team manager and what are the characteristics you should consider?

10. How would each of the team members from the various functional areas handle:

- Budgeting
- Capital acquisition
- Pricing

CHAPTER 7

Technology Transfer

1. INTRODUCTION

Technology transfer is the process by which technology, knowledge, and information developed in one enterprise for a particular purpose is applied and utilized in another enterprise for another purpose. Technology transfer can range from disseminating information on basic science research to commercialization of a specific product.

Knowledge varies, over time, from the initial concept of how a basic phenomenon can be applied to the solution of problems to knowledge applied to large complex systems. Technology transfer has been defined as the process of bringing both knowledge and application of science and engineering to society and to enterprises which do not employ this knowledge (Prehoda 1967, p. 87-88). The definition is incomplete, however, without a clear understanding of what technology transfer is designed to accomplish. Technology transfer can be categorized as scientific knowledge transfer, direct technology transfer, or spin-off technology transfer.

1.1 Scientific Knowledge Transfer
Scientific knowledge transfer is traditionally associated with research and development activities, and involves the transmission of knowledge gained through research. This type of transfer is most prevalent in the area of basic research and development. Knowledge transfer is accomplished primarily through information exchange and presentations of technical papers at scientific meetings and symposia.

1.2 Direct Technology Transfer
Direct technology transfer occurs between:
- Enterprise elements, i.e., internal integration

- Enterprise to enterprise
- Government to enterprise
- Government to Government

Direct technology transfer usually occurs through formal arrangements.

1.3   Spin-off Technology Transfer

Spin-off transfer occurs when technology developed by one enterprise, in one technical area, and usually for one purpose, is applied and used for a different purpose in a different technical area or market application than those foreseen at the time the research and development was originated. The United States government has introduced numerous programs and expended considerable resources to commercialize various technologies originally developed for specific governmental and military purposes.   Each type of technology transfer has a number of factors which must be considered. The advantages achieved through technology transfer also have associated risks and disadvantages (Rogers and Valente 1991). The transfer of technology can occur in any of the various stages of technological development and utilization (see Chapter One - *Technological Advancement and Competitive Advantage*). Table 7.1 shows the stages of the technology transfer process.

2.   METHODS OF TECHNOLOGY TRANSFER

There are a number of methods for transferring technology, both informal and formal.   The formal processes are those which use legal arrangements between the participants in the transfer process. The informal processes include the transfer of technology either with or without the transferee's knowledge or formal recognition that this knowledge is or has been transferred.

2.1   Informal Processes

The informal processes of transferring technology include technical information exchange through published matter, either printed or by electronic media, meetings, symposia and individual exchanges.   In this process the originator of the information makes the technical information available through professional meetings, journals, articles, electronic exchange, and informal meetings and personal communication. Technologists have sometimes inadvertently exchanged valuable and sensitive technical information which has led to competitors capturing economic rents that rightly belonged to the initiator of the information exchange. In some instances the knowledge of genetic code has been placed on-line in electronic media such as the Internet only to be patented by others. The process of training scientists in academic research institutions is another informal method of technology transfer.

Table 7.1
Stages of Technology Transfer Process

Stages	Who Carries it Out?	Definition
Basic research	Universities and government research labs Public enterprises	Original investigations conducted to advance scientific knowledge that do not have the specific objective of applying this knowledge to practical problems or products, processes or services
Applied research	Private enterprises Public enterprises	Scientific investigations that are intended to solve practical problems which can result in new technological products, processes or services
Development	Private enterprises Public enterprises	Process of placing a new concept into a form expected to meet the needs of potential users
Commercialization	Private enterprises	Process through which a technological innovation is converted into a commercial product to be sold by a private enterprise
Marketing	Private enterprises	Process by which a product is packaged, distributed, and sold by a private enterprise to its customers

(Source: After data contained in Rogers and Valente 1991, p. 106.)

Acquisition of critical technical personnel is another informal method of technology transfer. While intellectual property can be protected, it is very difficult to protect knowledge and skills which key technical personnel have encoded within them. While employee agreements can preclude direct competition, in practice it is very difficult to enforce, especially if the personnel move to another country, taking the knowledge with them.

2.2 Formal Processes

The formal processes of technology transfer are generally *process-embodied* or *person-embodied* transfer within an overarching organizational framework (Rogers and Valente 1991, p. 104). These processes include outright procurement of a technology through its sale, licensing or acquisition of the enterprise in which the technology is embedded. Another formal process is technology transfer through formal agreements between governments, enterprises, individuals, research entities such as laboratories, and academic institutions. In this instance, formal legal arrangements are made, such as joint ventures, R&D consortia, co-operative agreements and other legal instrumentalities.

3. INTERNAL TECHNOLOGY TRANSFER

In large enterprises, technology is not often developed, produced and introduced through only one organizational element. This is changing through the use of multifunctional teams (see Chapter Four - *Generation of Technology* and Chapter Six - *Enterprise Structure and Design*). However, the movement of a technology from an R&D team to manufacturing and then to marketing is a very complex process. The actual process of transferring technology within the enterprise involves a number of decisions (Garvin 1992, p. 326-328). One of the decisions is on timing, i.e., when the technology is ready to move from research and development to production and finally to market. The competing concerns for making this decision include customer needs, i.e., when is the technology at a stage to meet the customer requirements? The timing decision also depends on development of specifications and operating parameters that are sufficiently detailed to insure efficient, repeatable production. Timing is a function of preemption of competition. A late market introduction can have significant adverse effects on technology profitability; however, premature release can also have major adverse impacts. The timing of internal technology transfer is critical.

Location of technology transfer is a critical decision factor, which includes whether the new technology should reside in an existing or new portion of the enterprise. The location decision requires an understanding of the relationship of technology to R&D elements, i.e., should the technology be placed in a portion of the enterprise or a new element which is close in both physical proximity and/or culture to the R&D element? The organizational location decision must take into account the relationship of the technology to suppliers and customers, and the availability of technical and marketing skills of the organizational element who will be the responsible enterprise element. Other factors impacting location of the technology transfer process are the existing technology products and process requirements, and organizational learning, i.e., the need to increase the technical capability of the entire enterprise.

Deciding which staff members should be involved in the internal technology transfer process, according to Garvin (Garvin 1992, p. 327), is an element of the process. Multifunctional teams composed of the developers and receivers of the technology are the most effective in insuring a smooth transfer. Communication between the various enterprise elements involved in the internal technology transfer process is a critical factor in the decision process. It is important to develop effective communication methods for bridging the cultural, informational and geographic differences that often separate enterprise elements in an internal technology transfer process. Multifunctional teams with responsibilities throughout the entire project greatly enhance communication. Where multifunctional teams do not exist, then direct transfer of personnel and formal documentation can assist communication

The internal transfer of technology differs within an enterprise depending on the nature of the technology being transferred. The internal transfer, or integration of technology is different due to the complexity of the technology being transferred from R&D or applied basic science to the product or production stage of development (Iansiti 1995). Iansiti developed an information process framework, shown in Figure 7.1, for the development of products based on new innovative capabilities. Figure 7.1 shows the informational relationship between research, technology integration, and the product development process. The model serves as a framework for understanding the development process of new technology which has both a high level of complexity and component uncertainty.

The process and success of the internal integration or technology transfer are associated with the overall effectiveness of the total enterprise developmental process. The research indicates that more successful enterprises are characterized by a system focused on the integration process (Iansiti 1995). These enterprises emphasize the gathering of accurate information on how the technical factors would impact functionality and cost. This informational gathering is obtained prior to moving the technology from R&D to product development. The problems which may occur in later

236

stages of product development are directly linked to the decision in the critical internal technology integration or transfer stage.

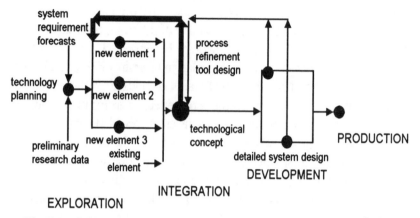

Fig. 7.1   Information Process Framework   (Source: Iansiti 1995, © 1995 IEEE. Reprinted by permission)

There are number of internal barriers to the transfer of technology between research and development and production (Vasconcellos 1994). One of the barriers is caused when R&D goals are not known by production managers. If production cannot stop to test new products and processes, this is a another barrier. When R&D is distant from *reality*, production will have little confidence in the transferred technology. Ineffective communication systems between production and R&D will form a barrier to the technology transfer. Another barrier is caused when R&D does not understand the needs and capability of production. The various levels of skills and expertise in R&D and production also add to the technology transfer barrier. Similarly, these factors will lead to production not trusting R&D due to the differences in culture between these elements. In general, production is resistant to innovation and is bound by routine, causing another barrier to effective technology transfer from R&D.

Vasconcellos' study indicated that the strongest barrier to internal technology transfer from R&D to production was the lack of effective communications (Vasconcellos 1994). The next strongest barrier was that the testing of new processes and products paralyzed the production line, especially in smaller enterprises. Again, one of the principal solutions is the utilization of multifunctional teams. Having production personnel participate with others in these multifunctional teams increases communication effectiveness and understanding of each of the participants' culture.

Vasconcellos' study of technology enterprises indicated that barriers could be significantly reduced by designing the technological strategic plan through the use of multifunctional teams. Decentralization of R&D, with these units closer to production, will also reduce the barriers. A better understanding of the culture of both enterprise elements can be achieved by internal position rotation between production and R&D. Vasconcellos also shows that linking and participation of marketing elements in the transfer process between R&D and production is also important since it offers insights into customer requirements.

4. EXTERNAL TECHNOLOGY TRANSFER

Successful external technology transfer depends on a number of factors including:

- Type
- Complexity
- Transfer mechanism
- Relationships
- Core competencies
- Organizational culture

According to Håkansson and Snehota "no enterprise is an island" (Håkansson and Snehota 1989). Technology external relationships are an important strategic consideration for amplifying internal capabilities and skills (Gemünden et al. 1994). One benefit is the synergy effects upon value and cost, i.e., increased value through increased quality and reduced cost through sharing resources and learning curves. The networking potential of external relationships, leading to cascading positive impacts, is another result of external technological relationships.

Figure 7.2 shows the relationships which interact with an enterprise. Gemünden et al. (Gemünden et al. 1994). Studies of four hundred and ninety-two enterprises showed a relevance of technology-oriented relationships to an enterprise's innovation capabilities. Figure 7.3 shows R&D intensity and co-operation as determinants of technological innovation success. Figure 7.3 illustrates that R&D external technology transfer contributes to reduced cost due to participating in R&D co-operation..

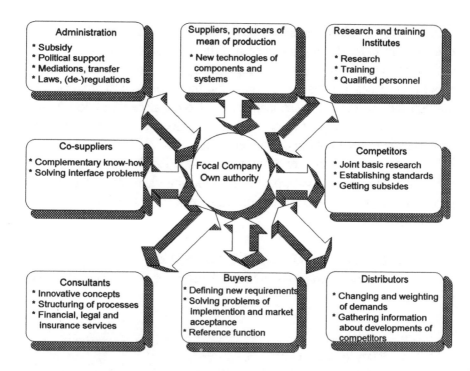

Fig. 7.2 External Enterprise Relationships (Source: Gemünden et al. 1994, © 1994 IEEE. Reprinted by permission.)

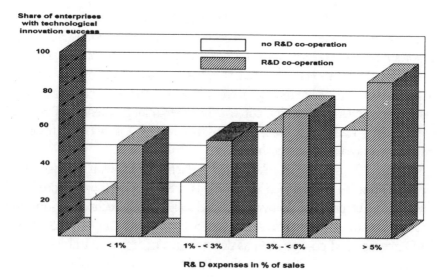

Fig. 7.3 R&D Intensity and Co-operation Impact (Source: Gemünden et al. 1994, © 1994 IEEE. Reprinted by permission.)

4.1   Methods of External Technology Transfer

An enterprise can acquire or transfer technology through a number of formal mechanisms which include:

- Co-operative and collaborative ventures
- Licensing
- Contracting
- Enterprise acquisition (see Chapter Four - *Generation of Technology*)

4.1.1 Co-operative and Collaborative Ventures

Co-operative and collaborative ventures are structures between two or more enterprises. These ventures can take various organizational forms (Tucci and Lojo 1994). One form of venture is through equity in a jointly owned new enterprise. A general partnership is another form of venture where the partners can be competitors, suppliers, or customers. The partners do not need to be limited to those within a single nation. Consortia consist of one or more enterprises including universities, industry and federal laboratories. Strategic alliances are an organizational format where two or more entities enter into a formal agreement to jointly pursue certain technological objectives. The term of the enterprise or partnership can be limited or unlimited in duration. The degree of sharing resources can vary from little to total, or be limited to specific aspects of the shared objectives. Collaboration, according to Tucci (Tucci and Lojo 1994), among participants can take the form of research, exchanging proven technologies across a single product line, joint development of one or more products, and collaboration across marketing and research.

The most frequently stated objective of co-operative and collaborative alliances is to develop or transfer new technologies (Tucci and Lojo 1994). A worldwide trend is the use of partnerships, strategic alliances, and other collaborative mechanisms to alleviate difficulties in all aspects of technological development (Bidault and Cummings 1994). These partnerships provide faster and less costly methods to develop new technology products, processes and services, e.g., SEMATECH, a collaboration of members of the United States semiconductor industry. Co-operative R&D allows partners to reach critical structures more rapidly for large complex technology projects, e.g., the European Airbus project. Partnerships and strategic alliances allow merging of technological knowledge and skills from multiple enterprises, improving innovation in the chosen technology. Bidault and Cummings (Bidault and Cummings 1994) found that cross-industry alliances proved more innovative than alliances with competitors. These cross-industry alliances occur more frequently than in the past. However, these alliances have a high risk of failure and produce more incremental than radical innovations.

Many characteristics of alliances are often in conflict with the objectives of the technology development originally envisioned. Technology driven co-operative and collaborative alliances can be managed through various enterprise structures. The strategic alliance all involve obtaining synergism through leveraging[1] mutual resources. This also achieves diversification of risk and mutual learning or technology transfer. Strategic alliances for technology transfer can be:

- Technical exchanges
- Cross-licensing
- Co-production
- Marketing agreements
- Joint product development programs
- Joint ventures with equity ownership

4.2   Intellectual Property
Another method for transferring technology is to transfer intellectual property, which is an intangible right that can be bought and sold, leased or rented, or otherwise transferred between parties in much the same way that rights to real property or other personal property can be transferred. Intellectual property can consist of patents, trade secrets, copyrights, designs, know-how, and trademarks. The transfer of intellectual property rights is an important and often substantial component of the technology transfer process. Intellectual property rights are most often transferred through contracts or licenses.

---

[1] Leveraging means obtaining more from combined resources than can be obtained by each entity separately.

4.2.1 Inventions and Patents

Invention is the act or process of discovering something new, physical or conceptual. A United States patent is an agreement between the United States government and the inventor. This agreement grants the inventor the right to exclude others from making, using, or selling the invention for a defined period of time within the United States. The patent law of the United States specifies that any person who

> "*invents or discovers any new and useful process, machine, manufacture, or composition of matter, or any new and useful improvements thereof, may obtain a patent.*"

A patent does not give the inventor the right to practice his or her invention, only the right to exclude others from doing so (von Hipple 1988, p. 47). The inventor is given exclusive use of the invention and the right to assign that use. However, a grant of a patent has been found not to be useful for excluding imitators and/or capturing royalty income in most industries. A grant of a patent is often likely to offer little benefit to its holder. Patents gives the patentee the right to exclude others from its use, but does not give the patentee the right to use the patent if such use infringes on patents of others. The United States patent system places the burden on the patentee to detect any infringers to sue for redress. A patent covers a particular means of achieving a given end, but not the end itself, even if the end and, perhaps the market it identifies, are novel.

4.2.2 Trade Secrets and Know-how

Trade secrets and know-how are others forms of intellectual property which can be used for technology transfer. A trade secret is any commercial formula, device, pattern, process, or information that affords an enterprise an advantage over others who do not know it. The information is not generally known and has value. Trade secrets must be maintained by avoiding public disclosure.

In contrast, know-how is a broader term that describes factual knowledge not usually amenable to a precise description. Know-how is usually accumulated knowledge as a result of trial and error. Know-how typically gives an enterprise the ability to produce something that could not be produced as accurately or successfully without it. Know-how may include trade secrets and cannot be protected or licensed unless it is first recorded in a tangible medium.

Unlike patents, in the U.S. trade secrets are protected by state, rather than federal, laws. These laws allow the trade secret owner to prosecute someone for unauthorized use or theft of such information. However, state laws generally require that a trade secret must be protected by its owner if it is to retain trade secret status. If the owner allows the information to become public information through publication, public use, observation, or lack of adequate security measures, the information moves into the public domain and loses protection under trade secret law. Trade secrets are effective to protect (von Hipple 1988, p. 54) product innovations that incorporate various technological barriers to analysis, and process innovations that can be hidden from exposure.

4.2.3 Copyrights

A copyright is an exclusive right granted by the United States government to authors, composers, artists, or their assignees for the life of the individual plus fifty years, to copy, exhibit, distribute, or perform their work. As with patent rights, these rights go to the individual creating the work, unless provisions are made to the contrary.

A copyright exists when a work is created. The law no longer requires the work to be marked with a copyright notice, but it is good practice to do so. Registration of copyrights with the federal government is optional and can be done at any time during the life of the copyright. Registration also permits using the federal court system to prosecute infringers and provides certain mandatory federal damages against those convicted of infringement. Registration may be recommended if software is the subject of a license agreement.

4.2.4 Trademarks

A trademark is a word, name, symbol, device, letter, numeral, or picture, or any combination of them, in any form or arrangement that is used to identify the origin of goods or services. A trademark must be individually identifiable and distinguishable from those of others for similar goods or services. Trademarks assure the buyer of the authenticity of a product or service and imply that the seller has exercised some standards of quality associated with the trademark. An enterprise or person may establish a trademark simply by using it in interstate commerce. Like copyrights, trademarks may also be registered.

4.2.5 Licensing

Licensing is the transfer of less than ownership rights in intellectual property to a third party, to permit the third party to use intellectual property. Licensing can be exclusive or non-exclusive, for a specific field of use, for a specific geographical area, or U.S. or foreign. If ownership is transferred, it is called an assignment.

The transfer of technology through licensing is a useful method for capturing economic rents of technological innovations. Small technology enterprises can benefit from technological licensing (Chung 1995). A small firm usually benefits more by licensing its technology than trying to commercialize it. The commercialization of new technologies requires a high expenditure of resources, generally beyond the means of many small entrepreneurial enterprises. Small enterprises, according to Chung (Chung 1995), to benefit from technological licensing, must overcome their initial naiveté by concentrating on the economic and strategic aspects of the process. These small enterprises must broaden operational perspective, scope, and build credibility and expertise. To further benefit from technological licensing, small enterprises should improve exchange and interaction capabilities, and enhance experience and responsibility.

5. INTERNATIONAL TECHNOLOGY TRANSFER

There are several differences between domestic and international technology transfer (Tucci and Lojo 1994). The level of competition in international technology transfer, i.e., the degree of substitutability of products, is lower than in domestic alliances, and it is more difficult to co-ordinate and transfer technology. International technology transfer is more difficult for enterprises outside the international alliance.

The benefits of lower competition outweigh the costs of linguistic and cultural differences. It is important that managers of technology consider the environment of the country associated with the international technology transfer (Edosomwan 1989, p. 95). Prior to developing long range technological plans, the enterprise can perform a detail study of the country. These studies should include a survey of financial institutions and instrumentalities that can be utilized in the other nation. The range of studies

244

performed includes a total enterprise assessment of the technology transfer project and contribution of the participants. Relationships with the governmental and private sector in the associated country also require development and nurturing. The project team and managers should be fully trained to deal with the various cultural, financial and legal issues that may arise in the international partnering. All the forms of technology transfer available between domestic enterprises are available in the international sector. The flows in international technology transfer are illustrated in Figure 7.4.

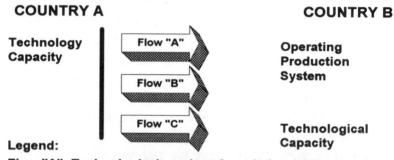

COUNTRY A   **COUNTRY B**

Technology Capacity   Flow "A"   **Operating Production System**

                                 Flow "B"

                                 Flow "C"   **Technological Capacity**

Legend:

Flow "A": Technological services & capital goods.
Flow "B": Skills & knowledge for operation and maintenance.
Flow "C": Knowledge, skills and experience to manipulate and change the producion system.

Fig. 7.4   International Technology Transfer (Source: Quazi 1995, © 1995 IEEE.
Reprinted by permission.)

6.   U.S. GOVERNMENT TECHNOLOGY TRANSFER

During the 1980s, the United States Congress began to focus on technology transfer as one way to mitigate the imbalance in international trade. The Congress was also concerned that United States leadership in science and technology was not translating directly into leadership in the private sector. A major gap emerged between the federal government's technology base and the private sector's willingness and ability to commit resources to applying these new technologies. The United States Congress and the Executive branch of the United States government firmly believes that effective transfer of government-developed technology could benefit society in:

- Developing a stronger domestic economy
- Creating new jobs
- Increasing the nation's standard of living

In 1986, government-owned, government-operated ("GOGO") laboratories were authorized to form a specific type of co-operative research and development agreement ("CRADA") with private sector partners. With the passage of the National Competitiveness Technology Transfer Act of 1989 ("NCTTA"), Congress mandated technology transfer as a mission for federal research and development laboratories and provided government-owned, contractor-operated ("GOCO") laboratories the same authority to employ the CRADA mechanism to promote industrial collaboration with contractor-operated laboratories.

Setting governmental objectives for technology transfer involves considering the goals of the government's research and development programs in the light of the public and national security interests. This can be rather difficult in that some objectives seem to conflict with one another. To conduct both governmental research and development programs and provide for technology transfer to the private sector requires consideration of a number of factors (DOE 1991). Enhancing United States economic competitiveness by promoting the commercial application of government-funded science and technology is a major consideration of United States governmental research and development programs. The dissemination of basic and fundamental research results has been one of the major factors of the technology transfer program of the United States government. This dissemination of technology information, however, has raised the question of intellectual property. Protecting and licensing commercially valuable information and intellectual property developed under government-funded research and development programs is a consideration in technology transfer. A great deal of the technology generated by the United States government has been related to national defense programs. A major objective in technology transfer by the United States government is protecting national security and controlling exports of sensitive technology. Other considerations include supporting the continued viability of key sectors that are critical to national security and economic growth, and meeting international obligations. The future United States national needs which underlie future technological strength through the fortifying the science and mathematics education foundations, is a another factor in the technology transfer programs of the United States government.

6.1 United States Governmental Technology Transfer Mechanisms

The technology transfer mechanisms available to the federal government include (DOE 1991):

- Conferences and Symposia
- Consulting
- Collegial Interchange
- Contracts
- Cost-Shared Contracts
- Co-operative R&D Agreement ("CRADA")
- Educational Grants and Awards
- Exchange Programs
- Grants and Co-operative Agreements
- Licensing
- Visitors
- Work for Others ("WFO")

Conferences and Symposia constitute an informal transfer mechanism utilizing meetings for the discussion and interchange of ideas, information, and covers topics of current and common technological interest to the participants.

The *consulting* form of transfer consists of providing advice or information through various mechanisms. One of these mechanisms is through consulting by government laboratories with full-cost recovery, including formal contractual arrangements between the parties involved, special intellectual property terms in favor of the sponsor, agreement that the work performed is non-competitive with the private sector and requires agency approval.

Another form of technology transfer is through *collegial interchange*, an informal, free exchange of information between colleagues, but it does not include exchange of trade secrets or restricted information. Private consulting by the private sector to the governmental laboratory under a contract or by government laboratory employees to the private sector under the individual's private contractual arrangements also results in technology transfer. Such arrangements require governmental laboratory approval that insures that no conflict of interests exists.

Contracts are an acquisition instrument between the federal government and a contractor. The contractor agrees to provide supplies or services to the government in return for a fee. This is often an effective tool for the government to promote and fund research and development that can be subsequently transferred to the private sector. One form of the contract instrumentality is *cost-shared contracts*. Cost-shared contracts include in-cash or in-kind arrangements. Patent rights are allocated in the same manner as other contracts. This type of collaboration must be of mutual benefit to both the private sector and the government, and must meet government requirements. Greater governmental control of the technical activity is maintained as compared to grants or co-operative agreements.

One form of technology transfer that has found favor in the United States is the use of *co-operative research and development agreements* ("CRADA"). A CRADA is an agreement between one or more United States governmental laboratories and one or more non-governmental entities under which the government, through its laboratories, provides personnel, facilities, equipment, or other resources with or without reimbursement. The non-governmental entities provide funds, personnel, services, facilities, equipment, or other resources to conduct specific research and development efforts that are consistent with the laboratory's mission.

Educational grants and awards are arrangements which provide an educational award, and the awardee retains all rights to the awardee's intellectual property. *Exchange programs* are short term agreements, usually less than one year, between a United States governmental laboratory and others to interchange information by the unilateral or bilateral exchange of personnel. The instrumentality of *grants and co-operative agreements* consists of assistance agreements that are entered into solely by the United States government with a recipient, whereby resources, money or property, are transferred to the recipient to support or stimulate research.

Using *licenses*, the U.S. government either licenses or acquires right through a license for a particular technology. *Visitors* are also an important part of technology transfer from and to federal laboratories. *Work for others* ("WFO") is any work performed by U.S. governmental laboratories for either the private sector or for other governmental agencies that use facilities or contractor personnel and are not directly funded wholly or in part by the sponsoring agency.

7.0   TECHNOLOGY ASSESSMENT

During the entire life cycle process of a technology, it is important to determine if the technological development is meeting, or will meet, the original objectives set prior to the decision to commit enterprise resources. This process of assessing the status of the technological development is not merely a process of observing a particular project or program but must be extended into the overall enterprise system for the delivery of technology. The technological delivery system which takes an innovation from conception through successful embeddment within its intended environment is the critical element for any technology. Many innovations have failed, not due to the technology but to the total system for translating the concept into a successful operational product, process or service. Therefore, it is important to assess the technology and the process of translation into a success entrant.

Technology assessment can occur in any of the stages of technological development (see Chapter One - *Technology Advancement and Competitive Advantage*). During the early stages, this assessment is performed to determine which technology offers the best opportunity to achieve the enterprise objectives which have been predetermined by the leadership team. In later stages of the process, the objective of the assessment is to determine if the technology under development is capable, or has a reasonable probability, of fulfilling the goals which have been set. After the development process has been completed and the technology has been introduced into its intended environment, the objective of the assessment is to review the entire process to learn if the process can be improved for future technological developments.

Assessing the technology delivery system is an important aspect of assessment. While changes can be made to a technological development as it is in the process of being translated into a product, process or service, it is the underlying system which may require reprocessing. Thus the assessment of the delivery process is key to future success in technology development. This process also requires that a member of the leadership team be assigned sole responsibility for insuring that technology is successfully developed, e.g., *Chief Technology Officer*.

7.1 Individual Technology Assessment

A technology must pass through three *gateways* to become commercially or socially embedded in its intended environment (Benson and Sage 1993). These *triple-gateways*, according to Benson and Sage, are:

- Market
- Systems-management
- Technology

A triple-gateway methodology, according to Benson and Sage (Benson and Sage 1993), is useful in the evaluation of emerging technology in the early outscoping phase prior to the expenditure of substantial enterprise resources. The outscoping phase is one of the seven phases in a multiphase life cycle analysis. A multiphase life cycle analysis is a process for determining the potential utility and costs of a technological development as it progresses from its conception through development and finally to embeddment or deployment. The multiphase life cycle analysis focuses on the early phases of the research and development process and includes scoping and identification of requirements specifications so as to identify societal and market needs (Benson and Sage 1993). Scoping includes the potential actuation process, associated technology issues, system management issues, and goals of the new technology. This requires documentation of technologies, economic needs, societal needs, and technological feasibility of the technological innovation under consideration. The Benson and Sage approach to technology assessment includes assessment and evaluation of alternative technologies, selection of appropriate technology, tracking progress of development and implementation, supporting the implementation of the technology, and disengaging once embeddment has occurred into the product development, manufacturing and market or curtailment. The triple-gateway assessment methodology starts with assessment of the market gateway.

7.1.1 Market Gateway
The market gateway assessment uses a market uncertainty analysis approach which covers new users, user skepticism behavior adjustment, competitive technologies, unpredictable technological developments, new uses, and legal barriers. The new user uncertainty derives from where a new technology is being offered to a user; an example is the use of Internet software to replace long distance voice communication and direct banking services.

User skepticism adds to technology uncertainty and results because developers of technology are usually optimistic about the relationship of performance and cost versus price. However, the users may not subscribe to the same set of relationships, i.e., the user may not want to pay the incremental pricing to obtain an incremental increase in performance. Uncertainty also arises due the need for user behavior adjustment. A new technological development may require users to change their behavior for the technology to achieve embeddment, e.g., telecommuting, where workers perform the majority of the work outside their enterprise offices.

New competitive technologies add to the complexity of the market adding considerable uncertainty, e.g., *JAVA,* a Internet computer language, which makes it possible to download software for one-time use, may cause existing operating systems to lose their importance. New technological breakthroughs also add uncertainty to the market. An example is the case of the early development of satellite communications which changed the international telecommunication industry structure. Uncertainty as to the market is added to by legal and regulatory requirements which can cause new technologies to be forced to overcome, for example, long periods of testing, adding to development cost and uncertain profitability.

7.1.2 Management Gateway

The management gateway is analyzed by reviewing the characteristics of the enterprise developing the technology (Benson and Sage 1993). This includes the enterprise modes of development:

- Entrepreneur
- Small high-technology enterprise
- Large enterprise with multiple markets and products.
- Multiple enterprises such as consortia and conglomerates which operate in multiple sectors.

Each mode incorporates risk avoidance, exploitation flexibility, marketing skills, legal skills, core competencies, and enterprise information and communication structure and culture.

7.1.3 Technology Gateway

The technology gateway is evaluated using uncertainty and risk analysis including technological innovativeness, and complexity due to technology, production and institutions (Benson and Sage 1993). Technological innovativeness uncertainty derives from the fact that the higher the degree of innovativeness of a new technological development, the larger the uncertainty. While Far Eastern enterprises use an incremental improvement approach, U.S. enterprises use a breakthrough approach with the commensurate uncertainty.

Technological complexity uncertainty appears to increase geometrically rather than arithmetically with the number of new technologies associated with a new technological development. An example is the considerable difficulty encountered in the development of the *Star Wars Project* of the United States Department of Defense, a program that no longer exists. Production complexity increases uncertainty, due to the difficulties in transforming a technology development from research and development to a product, process or service which can be embedded in its intended environment.

Institutional complexity increases market uncertainty since changes in the institutional structure can add serious complexity to a technological development. The introduction of *intelligent highways - smart roads* may be a future case where the institutional complexity raises questions as to the uncertainty of this technology. These triple gateways for assessing a technological development can be applied quantitatively and qualitatively as shown by Benson and Sage (Benson and Sage 1993).

8. ASSESSMENT OF INNOVATION

A technological development can be either a breakthrough or an incremental improvement. The type of development has, as previously been mentioned, major impacts as to development strategy and embeddment. The decision to develop either a breakthrough or an incremental improvement has major management implications. Each type of development has numerous subsidiary implications. It must be realized that few developments are truly breakthroughs, no matter what the developer may think; most are only minor incremental changes. In fact, a spectrum for innovation exists from breakthrough to minor or no change.

8.1 Technological Breakthrough

Understanding technological development and its degree of breakthrough or radicalness encompasses understanding the generation, development and use in its intended environment (Green et al. 1995). The degree of newness of development is one way of describing a technological breakthrough. How a technological development is perceived by the enterprise members, and the experience of staff, can serve as measures of the radical dimension of a technological development. The degree in which a development incorporates technology which is clearly a major departure from existing practice with associated risks can serve to define the degree of breakthrough.

Breakthroughs can be classified in terms of the degree of change made in existing enterprise practices.

Green et al. (Green et al. 1995) classify a breakthrough or radical technology by the extent to which the technological development incorporated embryonic and rapidly developing technology. The radicalness of a technology can be classified by the extent the development incorporates technology that is new to the enterprise but may be well understood outside the enterprise, and by how it departs from existing management or business practices. The amount of financial risk is another measure of the radical nature of the technological development. Green et al. (Green et al. 1995) showed through sampling a large number of research and development projects[2] that the dimensions of the degree of breakthrough or radicalness related to:

- Technological uncertainty
- Enterprise technological experience and knowledge
- Enterprise business experience and knowledge
- Fiscal resources associated with the particular technological development

9.   TECHNOLOGICAL METRICS

There are many different ways to view measures of technological performance. Performance metrics relate to who is viewing the technological development. Each viewer evaluates the performance based on his/her perspective. The United States Congress has a different viewpoint from a venture capitalist who is intending to invest in a particular development. These viewpoints can be summarized as:

- Societal
- Political or Governmental
- Enterprise
- System or Project

The metrics which are utilized by each of these viewpoints are quite different. A very positive metric from a societal viewpoint may from an enterprise viewpoint have a negative aspect, e.g., use of environmentally acceptable capital intensive technology for power stations. An example is a very low price for a drug which treats human immune deficiency virus

---

[2] 213 projects from 21 large industrial enterprises over four industrial sectors.

("HIV") versus a pharmaceutical enterprise requirement to recover high development and testing costs and achieve a suitable return for the equity owners.

9.1 Societal Metrics

The societal metrics for technological development derive from the fact that when a technology is introduced, extended, or modified it has consequences that are unintended, indirect, or delayed (Martino 1993, p. 10). The objective of societal metrics is to examine a potential technological development before it is deployed, since technology results in social change. There is no obvious deterministic link between technology and societal changes; however unanticipated and unintended societal changes have occurred because of the introduction of new technology.

The relationship of technological change to societal change is that it is a complex dynamical system which displays both periodicy and can lapse into chaotic behavior. The ability of complex dynamical systems to become chaotic is well documented and can arise from many sources (Çambel 1993, p. 117). An example is the competition of two different technologies and can be demonstrated by computations of the Lotka-Volterra equation (see Chapter Three - *Technological Forecasting*). Table 7.2 shows data on the impact of the introduction of the automobile. Table 7.3 illustrates that, while the societal costs can be large, the economic benefits for the automobile are also positive.

Table 7.2
Societal Impact of the Automobile

Year	No of Vehicles x 10³	Accidents x 10³	Deaths x 10³	State & Local Highway Debt Billion $	Insurance Premiums Billion $	Insurance Losses Paid Billion $
1900	8					
1910	3,000					
1920	8,131					
1930	23,035	9,859	32.9			
1940	27,465	10,333	34.5			
1950	40,333	10,418	34.8		2.6	1.1
1960	61,559	11,429	38.1	13.2	6.4	3.6
1970	89,200	22,116	54.8	19.1	14.6	11.2
1980	121,600	18,100	51.7	25.8	37.6	27.9
1990	150,000					

(Source: Based on data contained in Census 1990a ; Census 1990b)

Table 7.3
Economics of Automobile

	No. of Enterprises x 10³	Sales Billion $	Payroll Billion $	Employees Million
Dealers	103	333.4	28.7	1.37
Gas Stations	115	102.0	6.4	0.70
Repair	114.5	28.7	7.7	0.49
Rentals	11.4	16.4	2.4	0.13
Parking	9.3	2.6	0.5	0.05
Vehicle & Parts Manufacturing	4,422	204.7	22.9	0.75
TOTALS	4,775.2	687.8	68.6	3.49

(Source: Based on data contained in Census 1990b)

9.2 Political or Governmental Metrics

The United States Congress and the executive branches of the government have often viewed technological developments as a means to increase the nation's competitiveness (Spann et al. 1995). The value of governmental sponsored research and development to accomplish technology transfer is uncertain. Considerable legislative discussion in the Congress of the United States has done little to clarify these uncertainties. Table 7.4 shows a framework that Spann et al. (Spann et al. 1995) developed for studying technology transfer metrics. Spann et al. used three dimensions for their study of the metrics of technology.

Technology takes place over a span of time and the length of time can be short or long, and forms the *temporal dimension*. The temporal dimension is technology-specific and varies from technology transfer to transfer. Technological strategies can be either technology push or pull (see Chapter Two - *Technological Strategy*). Technology push strategies are used by the federal government laboratories. These strategies on the part of these laboratories result from their technological capabilities and are means-motivated. Technology pull strategies are demand-driven and needs-motivated. The technology push strategy focuses on counting metrics, such as the number of licenses issued and papers published. Technology pull strategy metrics rely on the level of effort expended and the level of resources committed to technology transfer activities. The particular technological transfer models used are based on the explicit or implicit transfer by managers of technology. These strategic models include those for technology push and pull. The technological push strategies include political models such as those used for justification of U.S. governmental R&D expenditures. Out-the-door technology push models are used for making technologies available for transfer, while opportunity cost models are employed when there is competition with other technological programs for resources.

Table 7.4
Technology Transfer Metrics

Temporal Dimension	Short Term	Long Term
Transfer Strategy	Technology Push	Technology Pull
Transfer Models	Political Model Out-the-Door Model Opportunity Cost Model	Market Impact Model Economic Impact Model

Measures	Licenses
	Request for Help
	Site Visits
	Tech Briefs/Paper
	  • Requested
	  • Published
	Technical Presentations
	Time Spent
	Transfer Budgets
	Transfer Expenditures
		Competitive Adv. Grants
		Cost Savings
		Jobs Created
		Market Share Gains
		New Businesses Started
		New Commercial Customers
		New Products
		Productivity Gains
		Royalties
		Return on Investment
		 Success Stories
		Technical Problems Solved
		User Satisfaction

(Source: Spann et al. 1995, © 1995 IEEE. Reprinted by permission.)

Technology pull strategies include market-impact models which are used for perceived needs to improve the global competitiveness of the United States and efforts to commercialize technologies developed at federal laboratories (Spann et al. 1995). Economic-impact models for technological pull strategies develop measures related to the perceived needs of the United States' economy, specific industries or individual enterprises.

The Spann et al. study found that metrics used to measure technology transfer were either sponsor- or developer-based. The technology push metrics used by the sponsor included:

- Requests for Help
- Licenses Granted
- Number of Site Visits

The technology pull metrics used by the sponsors of research and development included the number of jobs created, royalties collected, technical presentations made, success stories published, new businesses started, technical problems solved, number of new products developed, degree of user satisfaction, and the number of new commercial customers obtained. The developers used only one technology push metric, i.e., the number of technical presentations.

The studies of Spann et al. (Spann et al. 1995) also indicated other measures of technology transfer effectiveness. The metrics used as measures of technological transfer included:

- Technical Briefs/Papers Requested
- Technical Presentations
- Success Stories Published
- Technical Briefs/Papers Published
- Site Visits
- Requests for Help

The metrics for measuring commercial success included:

- Commercial Customers
- Commercial Sales
- Royalties
- Licenses Granted
- New Businesses Started

The measure of adopter benefits were found by Spann et al. (Spann et al. 1995) to be related to the number of:

- Technical Problems Solved
- Cost Savings
- Productivity Gains
- User Satisfaction

There is no one measure of technology transfer: the viewpoint of the measurer determines how success is measured. It is important that a shared understanding of the acceptable metrics important to each of the participants in the technology transfer process be developed (Spann et al. 1995).

9.3   Enterprise Metrics

An enterprise assessment viewpoint of a technological development is very different from either a societal or governmental viewpoint. An enterprise is usually concerned with how a new technology product, process or service can achieve market success. A new technological entrant into a market can create additional demand for the product and, or share the existing market, drawing buyers away from incumbent products, processes or services

(Mahajan et al. 1994). An example is the increased sales for all Microsoft® products with the introduction of Windows®95. The principal metrics for an enterprise are market share and financial performance

However, the addition of a new market entrant may expand the total market volume, because market entry is usually accompanied by an increase in product variety, promotional activity, reduced prices and increased distribution. Figure 7.5 shows a diffusion model for a new market entrant used by Mahajan et al. The parsimonious diffusion model approach determines the impact of competitive entry on market size and sales of the enterprise. Mahajan et al. (Mahajan et al. 1994) used this model to study the competing sales for the Polaroid Pack, Polaroid Integral, and Kodak Integral cameras.

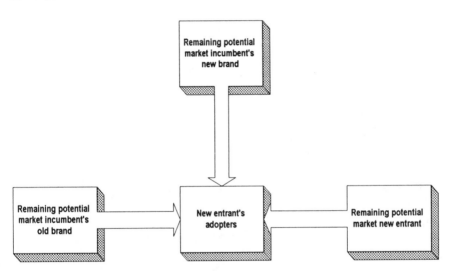

Fig. 7.5   New Entrant Diffusion Model (Source: Mahajan et al. 1994, © 1994 IEEE. Reprinted by permission.)

9.4   System or Project Metrics

Project metrics are related to technical performance measures and are quantitative design-related factors. A system cost-effective model (see Figure 7.6) approach has been proposed by Blanchard and Fabrycky (Blanchard and Fabrycky 1990, p. 80). Cost effectiveness relates to the metrics of a system in terms of system effectiveness. Cost effectiveness can be expressed as specific ratio metrics, such as:

258

$$\text{Benefits Effectiveness Metric} = \frac{\text{System Benefits}}{\text{Life - Cycle Cost}}$$

$$\text{System Effectiveness Metric} = \frac{\text{System Effectiveness}}{\text{Life - Cycle Cost}}$$

$$\text{Availability Effectiveness Metric} = \frac{\text{Availability}}{\text{Life - Cycle Cost}}$$

$$\text{Capacity Effectiveness Metric} = \frac{\text{System Capacity}}{\text{Life - Cycle Cost}}$$

$$\text{Supply Effectiveness Metric} = \frac{\text{Supply Effectiveness}}{\text{Life - Cycle Cost}}$$

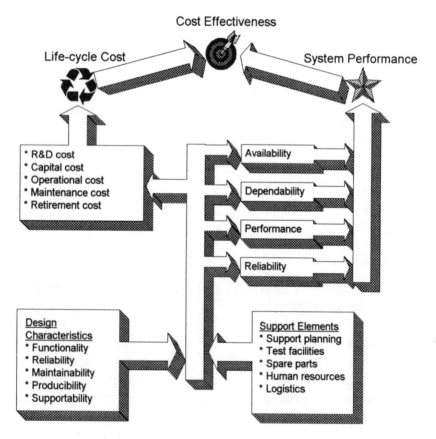

Fig. 7.6   Cost Effective Model for Projects after Blanchard and Fabrycky (Source: Blanchard and Fabrycky 1990, p. 80.)

Another means to measure system or project performance is through the technical performance measurement ("TPM") process (Archibald 1992, p. 283-287). TPM is a process to continuously predict or demonstrate the degree, anticipated and actual achievement, of technical objectives. The objective of TPM is to provide visibility of actual versus planned technical performance and early detection or prediction of problems. TPM serves as a means of determining the impact of the proposed alternatives.

TPM includes analysis of the differences in project performance in a time, performance and specification phase space. The *achievement to date* variable measures the value of specific technical parameters estimated or measured in a particular test. The *current estimate* tracks the value of specific technical parameters to be achieved at the end of the project if the current plan is followed. Tracking the progress of meeting the developmental specifications is another measure of system or project performance.

9.5   Technological Benchmarking

Technological benchmarking is the process by which an enterprise performs a direct comparison of its technological performance and other entities with similar operations. Technological metrics are an important element in the benchmarking process. The evaluations can be with competitors or others who are the best, or *world class*, in the specific technological areas in which the enterprise participates. The benchmarking process consists of ten steps (Camp 1989):

- Identification metrics for the benchmarking process
- Identification of enterprises to be compared
- Data methodology and collection
- Analysis of current performance and differences
- Forecast of future performance levels
- Communication of findings of the process
- Functional goal establishment
- Development of enterprise action plan to improve the performance in reference to comparative *world class* enterprises
- Implementation and monitoring of action plan
- Recalibration of the developed benchmarks to determine if improvement has been achieved.

World class entities are significant technological performers among their members on a total world basis. The compilation of *world class* enterprises in a specific technological area is not an easy task. In fact, entities outside a specific technological area may have a set of *best practices* which would define them as *world class* for that particular function. This process requires the expenditure of time and resources which may be difficult for some enterprises.

While benchmarking can be very useful, it has several enterprise barriers according to Cooper and Kleinschmidt (Cooper and Kleinschmidt 1995), which must be overcome. These barriers include the time required to accomplish the benchmarking process and co-operation of other enterprises to acquire comparison data. If an enterprise has not performed a benchmarking process before, the enterprise may experience difficulty in developing the correct methodology, i.e., metrics, sources, and interpretation. Another potential barrier for an enterprise in the benchmarking process is the link between best practices and improved performance, i.e., unknown underlying factors.

A study of one hundred and thirty-five enterprises by Cooper and Kleinschmidt (Cooper and Kleinschmidt 1995), to determine *best practices* as related to new product performance, used the following metrics:

- Sales Objective Success - How successful was the enterprise in meeting sales objectives previously set for a new market entrant?
- Profit Objectives Success - How successful was the enterprise in meeting profit objectives?
- Success Rate - The percentage of new innovations which achieved commercial success.
- Sales Impact - Impact of new product on overall enterprise sales.
- Relative Profitability - How the enterprise's new product program compares to similar programs of competitors or industry leaders?
- Overall Success - How on an overall basis does the enterprise compare to its competitors or industry leaders?

These metrics are converted to a one hundred point basis for a means of comparison. Figure 7.7 shows how similar benchmarking factors for technical support were presented by Colmen (Colmen 1994).

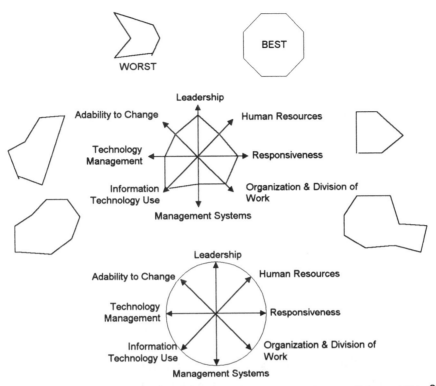

Fig. 7.7 Sample of Benchmarking Factor Performance (Source: Colmen 1994, © 1994 IEEE. Reprinted by permission.)

This type of research to develop benchmarking usually consists, according to Cooper and Kleinschmidt (Cooper and Kleinschmidt 1995), of a process defined by Colmen (Colmen 1994) and includes the development of the list of practices which define the critical success. The next step is the development of factors of performance metrics, i.e., measures are gauged via one to five Likert-type scales with anchor phases. Development of a questionnaire containing measures which captures the practices and performance is then prepared and administered to the enterprise being benchmarked. Prior to administration of the questionnaire, extensive pre-tests are conducted via personal interviews. The results of the Likert-type scales are converted to a one hundred point scale for analysis. In addition to the results of the questionnaire, other enterprise characteristics such as size, research and development spending, location and industry are obtained. Enterprises for comparison are then selected from private lists, compiled from data bases and directories of companies in the area to be benchmarked, i.e., new product development, innovation, and technical service. The questionnaire is directed to the comparative enterprise executives responsible for the area under study. To identify best practices, follow-up personal interviews are conducted with particularly proficient enterprises. The results of benchmarking studies are used for the preparation of action plans to improve the performance of the enterprise. An example of the type of best practices reported by Colmen (Colmen 1994) for benchmarking studies of technical support include:

- Active commitment of top management to technology in good times and bad.
- Insuring that the technical team is aware of corporate strategy, direction, and change.
- Special attention to long-term technical needs, including core competencies.
- Evaluating technical staff on performance rather than outcome, for higher risk projects.
- Flexibility in recruiting practice, providing for mid-level hires that add outside experience.
- Taking management risk to enter market prior to competitors.
- Decentralizing information technology function.
- Using complementary and supplementary external technical resources extensively.
- Continuous world-wide monitoring and assessment of critical impact technologies and associated reaction to threats and opportunities.
- Anticipation of environmental requirements and processes simultaneously for competitive advantage.

The acquisition of the *best practices*, which have successfully propelled other enterprises to a dominant or *world class* position in their respective fields, is one of principal results of a benchmarking study. When benchmarking is done for each of the major enterprise functions, the leadership team can determine where realistic improvements are possible. This type of analysis provides technology managers with improvement concepts that are external to the enterprise.

10. TECHNOLOGICAL AUDITS

The concept of audits has a long history stemming from the need to determine the validity of the financial system of the enterprise. Financial audits have historically been conducted by an outside impartial credited auditor. This concept, in the 1950s, was extended to the development of internal enterprise auditing practices. The internal audit function is defined as (Mints and Witt 1970):

> "*Internal auditing is an independent appraisal activity within an organization for the review of accounting, financial and other operations as a basis for service to management. It is a managerial control, which functions by measuring and evaluating the effectiveness of other controls.*"

The types of audits usually considered in an enterprise include financial, operational, organizational, functional, and management. A *technological audit* serves to review the adequacy and effectiveness of the technology management system within the enterprise. Technological audits pursue all the factors relevant to the successful development and introduction of new technology. The approach used is a total enterprise analysis. This means that the audit considers the various technological life cycle stages and determines which management factors are positive or negative for the successful completion of the entire process. Goodman and Lawless (Goodman and Lawless 1994, p. 119-133) discuss a two-stage *innovation audit*, a form of a technological audit, for technology enterprises.

The technological audit may reveal the need for additional in-depth technology delivery system studies and benchmarking. The technological audit reviews, analyses, and recommends. This audit is not involved with redesign or re-engineering of the technological delivery system of the enterprise. This maintains the objectivity of the process, which can supply the leadership team reliable information for managerial decisions affecting all phases of the technology delivery system of the enterprise.

The technological audit steps include:

- Familiarization with the technological objectives and operations of the enterprises.
- Examination of the technology delivery system used to achieve the technology objectives.

- Appraisal and evaluation of the adequacy and effectiveness of the total technology delivery system of the enterprise.
- Reporting of findings and constructive recommendations.

11. CHIEF TECHNOLOGY OFFICER

Many enterprises assign an individual of the leadership team to be the principal focal point for technology and for resolving the competing demands for technical skills and resources (Lewis and Linden 1990). This individual is usually given the title of *Chief Technology Officer* ("CTO"). Colmen (Colmen 1994) argued that the individual in the position should set priorities and decide strategic imperatives, customer and supplier interface procedures, environmental and safety requirements, new technological products, and processes and services for the entire enterprise. A modified set of Colmen's functions for a CTO consists of:

- developing the technological strategic plan
- developing and operating the technology environmental monitoring system
- serving as the Chief Technology Decision Maker ("CTDM")
- acting as the *technical conscience* of the enterprise
- providing technology oversight for the enterprise
- serving as the enterprise's Chief Technology Transfer Officer ("CTTO").

11.1 Technological Strategic Plan

The CTO interfaces with senior and technology enterprise leadership to develop technology objectives, goals and programs for the enterprise. The position guides the development of a technological strategic plan (see Chapter two - *Technological Strategy)* for the enterprise. The CTO recommends to the other members of the leadership team the steps necessary to implement the strategic plan in both the short and the long term. A survey that was conducted of technological enterprises in the Washington, D.C. Metropolitan Region of the United States indicated that over fifty-four percent of the respondents did have a technological strategic plan (Cardullo 1996).

11.2 Technological Environmental Monitoring System

The CTO develops and is responsible for tracking internal and external technological developments which can impact the enterprise. This includes identification and support of advanced and enabling technologies that are threats to, or opportunities for, the core enterprise. This can be achieved through a technological environmental monitoring system, which is a formal system for the collection and analysis of external and internal technology data including developments, markets, legislation and other factors which may impact the enterprise's technological developments (Martino 1993, p. 187-206). The survey of the technology organizations by Cardullo (Cardullo 1996) indicated that thirty-one percent of the survey's respondents had a formal technological environmental monitoring system.

11.3 Chief Technology Decision Maker

The CTO serves as the *master gatekeeper* for external technology which includes selecting and conducting long-term research consistent with the vision of the enterprise. This is not to imply that the CTO's role supersedes that of the Chief Executive Officer ("CEO"), but that the CTO is the principal technology adviser to the CEO or other Chief Decision Maker ("CDM") of the leadership team.

11.4 Technical Conscience

Colmen (Colmen 1994) states that *"A chief technology officer should be a statesman, conscience and traffic cop."* The CTO would, under this capacity, determine the balance between short-term and long-term technology programs. Included is assessing the programmatic needs to insure balance of financial resources between the program elements. Protecting the integrity of the research and development against the demands of technical services is also a CTO function.

11.5 Enterprise Technology Oversight

The CTO develops technology metrics for the enterprise, which includes performing technological benchmark studies. This oversight function includes the design and guidance of technology audits. Based on these activities, the CTO prepares reports on the status of the enterprise's technological fitness and make recommendations to the other members of the leadership team and the CDM. The CTO advises the leadership team about the technical implications of strategy, policy and programs (Betz 1993, p. 396).

11.6   Chief Technology Transfer Officer

One of the basic functions of a CTO is to promote internal enterprise technology transfer, serving as the interface for any external transfer, representing the enterprise at technical forums, committees and other appropriate activities. The technology transfer functional area would benefit the enterprise by the interaction of the CTO with major university, industry and government research consortia, and other groups associated with rapidly changing technology.

11.7   Leadership Team

The functions discussed above imply that a CTO should be an important member of the leadership team of the enterprise similar to the other members, i.e., Chief Executive Officer ("CEO"), Chief Operating Officer ("COO"), Chief Financial Officer ("CFO"), Chief Information Officer ("CIO"), Senior Vice Presidents, Vice Presidents and other senior enterprise executives. The size and members of a leadership team vary with the type and size of technology enterprise. In small and medium size enterprises many leadership functions are combined. Some technology enterprises which are primarily information-oriented combine the function of the CTO with that of the CIO. While the exact title is secondary, the functions of the various members and there inter-relationships are primary.

11.8   Enterprise View of CTO

A preliminary survey was designed by Cardullo (Cardullo 1996) to determine the manner in which technology enterprises use the CTO function. The objective of the survey was to determine the function and position of Chief Technology Officer as they currently exist within technology enterprises in the region surrounding Washington, D.C., the capital of the United States. While the formal assignment of this managerial responsibility is relatively new, some enterprises have been operating with an individual engaged in the function of CTO for considerable time. However, in some organizations this function may actually reside in another leadership team position. The questionnaire tried to capture how the enterprises dealt with these responsibilities.

The questionnaire focused on the nature of the leadership team of the various enterprises surveyed. This focus included the structure of the leadership team, use of the CTO function implicitly and explicitly, the development and use of technological strategic plans, and technological environmental monitoring systems. The questionnaire also sought an expression of opinion (Likert scale) with respect to the responsibilities of the CTO, regardless of how these might be assigned or performed within the enterprise. These opinions included:

1. Chief Technology Officer is vital to the growth and sustainability of an enterprise.
2. Chief Technology Officer should be the Chief Technology Decision Maker of an enterprise.
3. Chief Technology Officer should act as the *master gatekeeper* for external technology.
4. Chief Technology Officer should select and conduct long-term research consistent with the vision of the enterprise.
5. Chief Technology Officer should act as the technical conscience of the enterprise.
6. Chief Technology Officer should determine the balance between short-term and long-term programs.
7. Chief Technology Officer should assess programmatic needs to assure a balance of financial resources for the enterprise's technological developments.
8. Chief Technology Officer should design and guide technological audits.
9. Chief Technology Officer should be responsible for technological benchmark studies.
10. Chief Technology Officer should provide technology oversight for the enterprise.
11. Chief Technology Officer should serve as the enterprise's Chief Technology Transfer Officer.
12. Chief Technology Officer should promote internal enterprise technology transfer.
13. Chief Technology Officer should represent the enterprise at technical forums, committees and other appropriate activities.
14. Chief Technology Officer should be responsible for recruitment of both direct and indirect key personnel.
15. Chief Technology Officer should be responsible for and heavily involved in staff assessment, feedback and improvement.
16. Chief Technology Officer should serve as the enterprise's interface for any external technology transfer.

The preliminary survey, according to Cardullo (Cardullo 1996) of the Chief Technology Officer function indicated that the majority of the enterprises employed either a formal CTO or had another member of the leadership team filling that responsibility. In reference to the opinions requested, the majority of respondents either agreed or were neutral as to the functions presented. However, the oversight, conscience and benchmarking functions appeared to have a stronger favorable opinion than the recruitment function.

Each era brings changes to technology and enterprises. Enterprises must respond to those changes. The CTO is a response to the rapidity of technological development. The future of technology management requires that enterprises and the members of the leadership team seek innovative approaches not only to technological challengers, but to the management of technology.

REFERENCES

Archibald, R. D. (1992). *Managing High-Technology Programs and Projects*, John Wiley & Sons, Inc., New York, NY.

Benson, B., and Sage, A. P. (1993). "Emerging Technology - Evaluation Methodology: with Application to Micro-electromechanical Systems." *IEEE Transactions on Engineering Management*, 40(May), 114 - 123.

Betz, F. (1993). *Strategic Technology Management*, McGraw Hill, Inc., New York, NY.

Bidault, F., and Cummings, T. (1994). "Innovating Through Alliances: Expectations and Limitations." *IEEE Engineering Management Review*, 22(Fall), 116 - 124.

Blanchard, B. S., and Fabrycky, W. J. (1990). *Systems Engineering and Analysis*, Prentice-Hall, Englewood Cliffs, NJ.

Çambel, A. B. (1993). *Applied Chaos Theory: A Paradigm for Complexity*, Academic Press, Inc., Boston, MA.

Camp, R. C. (1989). *Benchmarking: The Search for Industry Best Practices that Lead to Superior Performance*, ASQC Quality Press, Milwaukee, WI.

Cardullo, M. W. "Chief Technology Officer: A New Member of the Leadership Team." *UNESCO Conference on Managing Technology*, Istanbul, Turkey.

Census. (1990a). *Historical Statistics of the United States*, U.S. Bureau of the Census, Washington, D.C.

Census. (1990b). *Statistic Abstracts of the United States*, U.S. Bureau of the Census, Washington, D.C.

Chung, K. B. "Technology Licensing for the Small Firm." *IEEE Annual International Engineering Management Conference*, Singapore, 309 - 314.

Colmen, K. S. (1994). "Benchmarking the Delivery of Technical Support." *IEEE Engineering Management Review*, 22(Winter 1994), 47 - 51.

Cooper, R. G., and Kleinschmidt, E. J. (1995). "Benchmarking Firms' New Product Performance and Practices." *IEEE Engineering Management Review*, 23(Fall), 112-120.

DOE. (1991). "Technology Transfer: A DOE and Industry Partnership for the Future.", U.S. Department of Energy, Washington, DC.

Edosomwan, J. A. (1989). *Integrating Innovation and Technology Management*, John Wiley & Sons, New York, NY.

Garvin, D. A. (1992). *Operation Strategy: Text and Cases*, Prentice Hall, Englewood Cliffs, NJ.

Gemünden, H. G., Heydebreck, P., and Heider, R. (1994). "Technological Interweavement: A Means of Achieving Innovation Success." *IEEE Engineering Management Review*, 22(Summer), 48 - 58.

Goodman, R. A., and Lawless, M. W. (1994). *Technology and Strategy: Conceptual Models and Diagnostics*, Oxford University Press, New York, NY.

Green, S. G., Gavin, M. B., and Aiman-Smith, L. (1995). "Assessing a Multidimensional Measure of Radical Technological Innovation." *IEEE Transactions on Engineering Management*, 42(August), 203 - 214.

Håkansson, H., and Snehota. (1989). "No business is an island." *Scandinavian Journal of Management*, 5, 187 - 200.

Iansiti, M. (1995). "Technology Development and Integration: An Empirical Study of the Interaction Between Applied Science and Product Development." *IEEE Transactions in Engineering Management*, 42(August), 259 - 269.

Lewis, W. W., and Linden, L. H. (1990). "A New Mission for Corporate Technology." Sloan Management Review, 57-65.

Mahajan, V., Sharma, Subhash, and Buzzel. (1994). "Assessing the Impact of Competitive Entry on Market Expansion and Incumbent Sales." *IEEE Engineering Management Review*, 22(Spring), 13 - 23.

Martino, J. P. (1993). *Technological Forecasting for Decision Making*, McGraw-Hill, Inc., New York, N.Y.

Mints, F. E., and Witt, H. (1970). "Internal Auditing." Financial Executive's Handbook, R. F. Varcil, ed., Dow Jones-Irwin, Homewood, IL.

Prehoda, R. W. (1967). *Designing The Future - The Role of Technological Forecasting*, Chilton Book Co., Philadelphia, PA.

Quazi, H. A. "Application of TQM Principles in International Technology Transfer Process: an Integrating Framework." *IEEE Annual International Engineering Management*, Singapore, 128-133.

Rogers, E. M., and Valente, T. W. (1991). "Technology Transfer in High-Technology Industries." Technology Transfer in International Business, T. Agmon and M. A. von Glinow, eds., Oxford University Press, New York, NY, 103-120.

Spann, M. S., Adams, M., and Souder, W. E. (1995). "Measures of Technology Transfer Effectiveness: Key Dimensions and Differences in Their Use by Sponsors, Developers and Adopters." *IEEE Transactions on Engineering Management*, 42(February), 19 - 29.

Tucci, C. L., and Lojo, M. P. (1994). "Social comparisons and co-operative R&D ventures: The double edge sword of communication." *Journal of Engineering Management*, 11(December), 187 - 202.

Vasconcellos, E. (1994). "Improving the R&D-Production Interface in Industrial Companies." *IEEE Transactions on Engineering Management*, 41(August), 315 - 321.

von Hipple, E. (1988). *The Sources of Innovation*, Oxford University Press, New York, NY.

DISCUSSION QUESTIONS

1. Comment upon cost-sharing consortia with regard to industry/government co-operation, including advantages and disadvantages.

2. Describe and discuss what type of technology transfer mechanism you would use if you were:
   - Small entrepreneurial enterprise.
   - Multinational organization.
   - Government laboratory.
   - Government agency.

   Your discussion should include the advantages and disadvantages as viewed by the various participants in the transfer.

3. Choose an instance of technology transfer either from your readings, handouts or experience and discuss lessons learned about the form of technology transfer used.

4. What mechanisms should be used to encourage technology transfer between individuals or units in an enterprise?

5. Define a set of technological metrics for assessing and benchmarking the following:
   - Scientific development with little or no potential commercial application.
   - Technological development for a purely government program.
   - Technological development for a rapidly evolving market.
   - Technological process for a mature industry.
   - Technological service for commercial introduction.

6. Survey literature, both electronic and standard media, for references to a Chief Technology Officer. Report your findings, summarizing the contents of the literature.

7. Review literature, both electronic and standard media, for references to Technological Audits. Report your findings, summarizing the contents of the literature.

INDEX

N

O

U